THE HUTCHINSON LIBRARY OF SHIPS AND SHIPPING

A History of Marine Engineering

THE HUTCHINSON LIBRARY OF SHIPS AND SHIPPING

General Editor: A. Silverleaf
Director, Road Research Laboratory; previously
Deputy Director, National Physical Laboratory

ALREADY PUBLISHED

Basic Naval Architecture K. C. Barnaby
Some Ship Disasters and their Causes K. C. Barnaby
Design of Marine Screw Propellers T. P. O'Brien
Merchant Ship Design R. Munro-Smith
Merchant Ships and Shipping R. Munro-Smith
Screw Tug Design J. B. Caldwell and Jeffrey N. Wood
Bizarre Ships of the Nineteenth Century John Guthrie

John Guthrie

A History of Marine Engineering

Hutchinson of London

HUTCHINSON EDUCATIONAL LTD
3 Fitzroy Square, London W1

London Melbourne Sydney Auckland
Wellington Johannesburg Cape Town
and agencies throughout the world

First published 1971

This book has been set in Garamond type, printed in Great Britain
on cartridge paper by Anchor Press, and
bound by Wm. Brendon, both of Tiptree, Essex

ISBN 0 09 108200 5

Contents

Illustrations

Acknowledgements

The three most important features in any history are accuracy, continuity and interest, and none of these can be obtained without good references. In the case of a technical history, these are liberally supplied by contemporary engineering periodicals and by the transactions of the appropriate engineering institutions, ably helped out in a practical way by our national museums. I am accordingly deeply grateful to the Editor of *The Engineer* and to the Secretaries and Librarians of the Institution of Mechanical Engineers, the Institute of Marine Engineers and the Royal Institution of Naval Architects for permission to use material from their magazines and proceedings. I am also indebted to the Trustees of the Science Museum, London, the National Maritime Museum, Greenwich, and to Nottingham Water Board for permission to reproduce photographs to illustrate the text, and to the Science Museum Library to make use of material in the Goodrich Papers.

In dealing with the more modern sections of the book: Gas Turbines and High-speed Craft, I have drawn heavily on the experience of a number of colleagues and friends, all experts in their subjects, among whom I would like to include Messrs R. J. Hook, A. K. Buckle, C. Beason. A. Silverleaf, P. R. Crewe, R. M. Duggan and A. T. O. Howell, while I am grateful to Messrs Ruston Gas Turbines Ltd., Lincoln, for their permission to reproduce a section through one of their turbines.

I must also pay tribute to the large number of friends and colleagues who, over many years, have helped in producing subject matter to develop my thesis, or drawn my attention to fresh lines of research; they are too numerous to name individually but are now thanked collectively.

Finally, I must not forget the grizzled stalwarts of my early days, engineers who had tended those giant engines on the Atlantic Ferry, and who could talk of the *Great Eastern* from personal experience—I now realise that their marvellous tales were not all fiction.

John Guthrie

Editor's Note

This series of books is primarily intended to be of interest to those professionally concerned with the design, construction and operation of ships and other marine vehicles. Many remarkable changes are now taking place in the size, shape, speed and capability of conventional ships of all types, while hovercraft, hydrofoil ships and other unusual vessels are beginning to have a striking effect on the maritime scene. Technical staff and management increasingly need up-to-date design data and specialist information on a wide range of topics, and it is hoped that most books in the series will be of direct value to them, and to many students at universities and technical colleges.

In addition to specialist monographs and student textbooks, the series also includes books having a broad appeal to all those who want to know more about the fascinating variety of craft which can be seen in ports, on rivers, and at sea: this book is one of that group. Its principal purpose is to make us aware of the way in which modern marine engines have been developed during less than two centuries from relatively crude and simple adaptations of the earliest steam pumping engines used in mines to the variety of powerful, compact, and highly reliable steam and gas turbines and diesel engines which serve at sea to-day. While a good deal of this history has been recorded elsewhere, most other accounts tend to be piecemeal and few, if any, combine the detail knowledge, continuing enthusiasm, and vivid direct personal impressions displayed here. Together these give the story a particular flavour which adds to its professional value as a coherent account of the remarkable advances made over the years by able and dedicated engineers both ashore and afloat, and which have transformed sea transport from an activity much dependent on the vagaries of wind and waves to a planned operation run to closely regulated timetables. The effects of this on the modern world are immense and still growing; indeed, the salt water engineer could claim to be one of the pillars of modern society if he were not far too modest to do so. These far-reaching consequences are implicit in this 'history book on marine engines', which describes not only their anatomy but also their idiosyncrasies and faults, and which will evoke many memories in all who have worked Behind the Iron Door, and will instruct and entertain everyone who wants to enlarge his knowledge of an exciting aspect of man's conquest of his natural environment.

Introduction—behind the iron door

The introduction to a book must not only attract the reader's attention, it must also give rather more than a hint of what it proposes to tell him. This aphorism may not be pertinent in the case of a textbook whose content has to be digested as a matter of discipline, but it is important in a volume written with a view to enlarging the reader's existing field of knowledge—not teaching so much as filling in the gaps. The iron door is evocative to any marine engineer: its loud clang at a quarter to midnight reminds him that in exactly thirty seconds his cabin light will be switched on and a hoarse voice remind him that it has gone one bell, time to get out of his bunk and go down on watch again. Sometimes the pattern changes and the door clangs at the wrong time, but the summons is still much the same: 'Everybody below, we're changing No. 6 piston', just the start of another long spell of unremitting labour. There may be more than one iron door on a ship, but *the* iron door is that leading from the engineers' alleyway into the engine room.

As for the title, 'history' is the only term I can think of that will apply in this instance—although the expression brings to mind the history lessons of my youth, sitting with glazed eyes in a stuffy classroom in high summer listening to a bored teacher droning on about some improbable treaty entered into long ago by impossible people. As a schoolboy the history taught to me dealt only with a very limited section of the community, with political agreements or wars between nations; it never mentioned the common man or described how the majority of mankind managed to survive throughout the millennia of developing civilisation. And that is just what conventional history cannot tell us. Until the Renaissance, not only the man in the street but his overlord and sometimes his monarch were illiterate, and the clergy were the only people who could undertake the task of recording events as they happened. But these clerics were not interested in the common man.

It occurred to me then, that a book written about ordinary people and how they existed under different rulers throughout the ages could hardly avoid being a best seller and would probably lead to a truer understanding of history than many a conventional textbook.

And so with marine engineering. I have read and enjoyed all the books on marine engine development that have come my way, yet invariably I have been disappointed with some part or other of the subject matter. Halfway through the books one sensed the story was being told at second hand because the authors were historians without practical experience. As in the school text books, nobody ever thought of explaining the more mundane facts of engineering history, such as how the early engines were built before the age of machine tools; what they were like to handle and how they behaved at sea, or how did one ascertain boiler water-level and pressure before gauges were invented?

Now a book recording the development of marine engineering over a period must not only describe the anatomy of the engines but also their ailments, their idiosyncrasies and their general cussedness in off-peak moments. This can be done effectively only by one who has lived with them, one who has long since learned that a prime mover is not just a complicated piece of ironmongery but is an entity with a personality of its own. And only he would appreciate the hidden meaning in such random remarks as are heard every day in our ports and docks, usually on sailing day after a hasty refit:—'Don't worry, Chief, the cylinder joints will take up before you get to Tilbury' or 'Those exhaust valves will have to do another trip, Second, just manage as best you can.'

Ships' engines do not always live up to builders' expectations; like human beings they have their seamy sides well known to the people who attend to their maintenance and repair, and again, they are in the main predictable. Most steam engines and all diesel engines conform to definite patterns of behaviour, and the understanding of these patterns is part of a classification surveyor's stock in trade. In a lifetime of work on thousands of ships, continually recording the weaknesses of their component parts, he eventually acquires an encyclopedic knowledge of all engines both ancient and modern, good, bad and downright awful. He is not taken in by builders' claims for near perfection, he has crossed the North Atlantic too often with their engines, but then neither is he fooled by the Chief Engineer, sobbing on his shoulder and asking for a reconditioned engine and new boilers.

Consequently, reverting to my earlier musings, I decided to write my own history of marine engines, describing not only their anatomy but also their vagaries and faults. In passing, I ask the reader's indulgence for vagaries of style and the use of engineer's shorthand: for example, I refer throughout the book to boiler pressure in lb rather than lb/sq in or psi. No historian can be expected to write from first-hand experience but at least he should expect to be primed in first-hand knowledge, and in marine engineering as with ships in general it is still possible to recapture the atmosphere of the nineteenth century.

In the 1920's, when I started to take an active interest in practical engineering, the earliest motorships were hardly in their prime and paddle steamers with delightfully crazy engines were thick underfoot. Passenger ships on the Atlantic Ferry were still being propelled by steam engines the size of cathedrals and it was possible to be completely lost in their multiple boiler rooms. Even today one can occasionally meet with a relic of the early age of steam in some sun-bleached

Central American port; that small graceful coaster with the clipper bow, the peeling white paint and the rakish yellow funnel, and with a quaint and ornate little engine that rightly belongs in a museum.

In dealing with ships' propulsion I have confined my history to that of the Merchant Navy—the Royal Navy's engines must await their own historian. Although the marine engine performs precisely the same function in both Naval and Merchant ships, the conditions are completely different and warrant separate treatment.

Finally, I shall attempt to provide answers to all those niggling little questions about the how and the why that suggest themselves in reading a conventional history and that are so frequently ignored by the Author, although I regret I have not yet discovered the meaning of a cryptic remark overheard in the engineers' alleyway in a large passenger steamer before the War:—'As long as that bit of wire holds, those boilers are safe as houses.'

1 The early steam engine

Although the history of ships and shipbuilding is as old as civilisation, the story of marine engineering goes back a bare 150 years, and most of that is the story of steam. It traces the rapid development of the steam engine, starting with the crude atmospheric engine, through compound expansion to high-speed rotary engines, using in turn the various characteristics of steam to create power by forming a vacuum through condensation, by exerting pressure on a piston or by kinetic energy through a turbine nozzle.

Present-day merchant ships are propelled exclusively by one of three distinct types of prime mover: the oil engine, the steam turbine or the reciprocating steam engine. The first is a comparative newcomer, a brash youngster with a brilliant future and the world at its feet; the second has by now reached a respectable middle age and an assured position, while the third is heading for a not-too-distant scrap-heap. The very few exceptions to the rule are the free-piston gas-turbine-propelled ships, whose machinery is a by-product of the oil engine, and the occasional ice breakers driven by nuclear power, in essence simply a steam turbine with a special type of boiler.

The marine steam engine, in one form or another, has been with us since the Napoleonic Wars, quite a long time when judged by human endeavour, but a woefully short period indeed in terms of technical progress. The art of working in metals is as old as war itself: iron swords were being forged 3000 years ago, while beautifully wrought and inlaid plate armour was being produced during the Middle Ages. Delicate rack-and-pinion winding gear was being fashioned for sixteenth-century crossbows, while in the eighteenth century precision mechanisms were fitted to small arms which for workmanship and finish put to shame the clumsy, uncouth valve motions fitted to contemporary steam engines. In the early nineteenth century rifled barrels were fitted to guns at a time when a steam cylinder could not even be bored reasonably true.

Why were the skill of the armourer and the art of the gold and silversmiths not applied to the new craft of engine-building? There is no short answer to this question. Plate armour and guns were made by craftsmen for princes and nobles, who alone could afford the price, while steam engines were built by artisans for mine-owners and industrialists of modest means who were not concerned with appearances, only with results. Also, the scale effect enters into the picture: small items could readily be shaped and polished at an instrument-makers' vise-bench where a heavy crankshaft could only be rough-forged to an approximate shape—the machine tool had still to arrive.

Hero of Alexandria was the first to demonstrate the possibilities of steam under pressure with his aeolipile in 130 BC, but it took over eighteen centuries

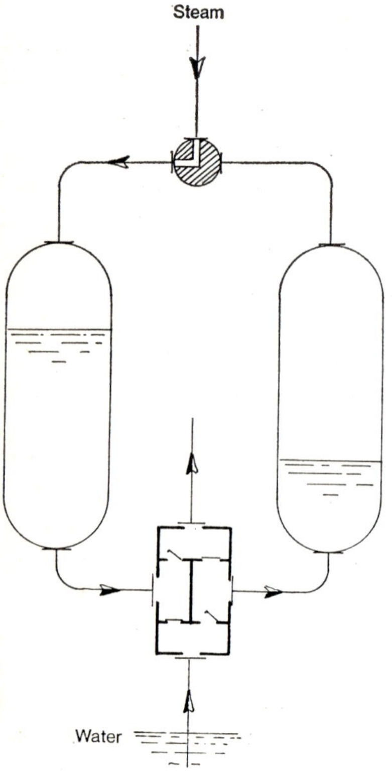

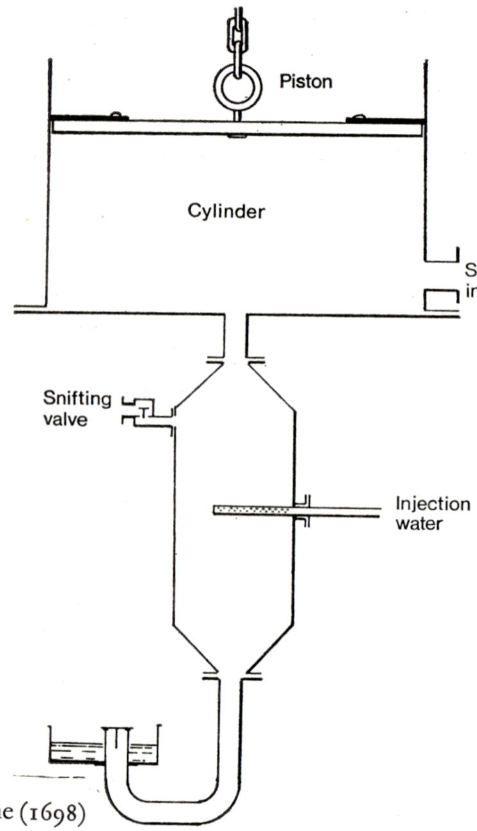

1 *Left:* Thomas Savery's pumping engine (1698)
2 *Right:* Picklepot condenser for a Newcomen engine, introduced about 1770

for the concept behind Hero's rotating toy to evolve into the steam engine.

It was not until the end of the seventeenth century that an expanding population and a more practical approach to mine operation offered suitable conditions for a form of cheap power. The difficulty of pumping out even shallow mines with teams of horses and even men, as used by the Germans in the sixteenth century, created a climate favourable to the invention of the steam engine.

The first man to give practical proof of the use of steam for converting heat into useful work was Thomas Savery, who in 1698 patented a machine for raising water from a mine sump and delivering it at a head of up to 80 feet. This machine consisted of a pair of metal receivers each with a steam inlet at the top and a water inlet at the bottom. The two steam pipes were connected through a three-way cock to a common steam pipe, while each water pipe was connected through a separate valve chest to a common suction and delivery pipe. The action of the pump was as follows: steam was admitted to receiver A and then shut off. As it condensed, water entered the receiver from the suction pipe through the valve chest. Upon steam being readmitted to the receiver, the water contained in it was forced out through the valve chest into the discharge pipe, the steam remaining in the chamber then condensing to repeat the cycle. A similar cycle was performed in receiver B, the two receivers working in parallel, the one drawing in a charge of water on the suction stroke while the other delivered its contents under pressure to the discharge line. No valve gear was required to work this pump, other than the switching over of the three-way steam cock at the end of each stroke, as the water chests were fitted with clack valves.

Through careful manipulation of the steam cock by the engine attendant a fairly smooth flow could be obtained, and according to Savery the timing of the stroke was judged by the appearance of the receivers under working conditions. The metal walls were dry and hot to the touch where the steam was, and covered in dew in way of the water, and the water-level could thus be confidently estimated. By careful attention to the rate of discharge, the operator could work his pump using the minimum amount of steam. A similar phenomenon, incidentally, is commonly observed in the CO_2 bottles when charging up a refrigerator.

The Savery pump was extensively used in mines and eventually became known as the Miners' Friend, although the delivery head was limited by the boiler pressure and the strength of the receivers. As would be expected of a machine where the pressure cylinder was also the condenser, the fuel consumption was extremely high: about 30 lb of coal per HP as near as can now be judged, or about twenty times that of a modern steam pump.

Savery's principle, however, is still used in the pulsometer-type pump where the two receivers with their valves are combined in the one casing, with a nicely balanced ball valve with two seats, one to each receiver, taking the place of the three-way steam cock. As the condensed steam is discharged together with the pump water and is thus not recoverable for re-use in the boiler, its general application is restricted to land installations, though marine engineers may be surprised to learn that it has not been totally excluded from shipboard use. In my early days as a classification surveyor it was my job to put an old-time coal-burning passenger ship through successive surveys and on each occasion her pulsometer-type general-service pump, about the size and shape of a large steamer trunk, was opened out, cleaned, examined and boxed up again without requiring any repairs whatever, except, possibly, new water-valve rubbers. This pump was supplied with a 3-inch-bore steam pipe and according to the admiring Second Engineer was the fastest pump on bilges he had ever come across, hinting darkly that anyone opening a strumbox with the pump working did so at his peril.

Thomas Newcomen, a blacksmith living at Dartmouth, and contractor to Thomas Savery, recognised both the possibilities and the limitations of the mine pump. He accepted the tremendous power inherent in a vacuum but rethought the engine by keeping the steam and pump ends separate: he designed an atmospheric engine that would act as a simple handle to lift a plunger pump, and therein lay his genius. This machine, the so-called Fire Engine, was patented by him in 1705 and the first pump was installed and set to work in 1712. It depended for its power upon the atmospheric pressure acting on top of a piston with a vacuum below it, and the action of the steam was merely to create this vacuum. The steam, in fact, was simply water vapour.

The engine, as originally designed, consisted of an overhead wooden beam with arched (horse's head) ends, oscillating on a supporting wall. A metal plate piston suspended by a chain from one end of the beam worked in a vertical open-ended brass cylinder set on top of the boiler, with a throttle valve between the two. At the other end of the beam hung the heavy pump rod with its plunger, and the engine stroke was limited by spring buffers acting on the main beam. The piston was made as good a fit as possible in the rough bore of the cylinder, and was provided with a leather sealing flap, covered by water from an overhead tank charged by the main pump. The water seal above the piston was essential, the supply pipe being up to 2 inches bore, and the depth of seal was maintained at 6–12 inches. The steam throttle valve was opened and closed by a plug rod or valve spindle worked from the beam, and a small cock on the bottom of the cylinder admitted a water spray for condensing.

The boiler was a simple tank, lightly built for minimal steam pressure as the engine worked on a vacuum, and the boiler manhole could be removed for feeding whilst under working conditions.

The action was very simple: due to the weight of the pump rod, the engine at rest remained with the piston 'outdoors', i.e. at the top of the stroke. Steam was admitted to the cylinder through the throttle valve by manual control to start with, then closed and the injection cock opened to create the vacuum and draw the piston down, thus raising the pump rod on the suction stroke. When the steam was again admitted the piston was allowed to rise to the top and the weighted pump rod made the working stroke on the pump. From this it will be seen that the pump was, in effect, worked by the weight of the pumping gear, and that the steam working cylinder acted merely to lift the plunger for a fresh stroke. The pump delivery head was now no longer a function of the boiler pressure, but of the weight of the pumping rods, and while there was a limit to these, mines up to 500 feet deep could be pumped out in two lifts.

Almost from the start, in the early 1700's, this pump was made self-acting, there being two operations to control the piston movement: the steam-throttle valve worked by a plug rod from the beam and the condensing spray cock opened by a buoy floating in a standpipe on the boiler. When the steam pressure increased, the water-level in the pipe rose, lifted the buoy, opened the cock, and thus started the piston on the down stroke, followed immediately afterwards by steam admission to the cylinder with consequent lowering of the boiler pressure. Incidentally, this buoy ensured that the engine consumed the steam as fast as it was generated. To get more engine strokes meant so much harder firing.

The myth of the boy attendant who adapted automatic trip gear to the Newcomen engine is now seen in its proper perspective—complicated and costly pumping engines were not left in the care of irresponsible boys.

The buoy control for the injection cock was very soon replaced by the cataract trip, a most reliable regulator and which suited the beam pumping engine to perfection, giving time for the pump chamber to fill before starting the working stroke. This cataract consisted in principle of a simple tipping bucket linked to the injection cock handle, and filled through a regulating cock from the overhead tank. When the water reached a predetermined level the bucket tipped over, emptied and immediately righted itself, and this action opened the injection cock for a brief interval, sufficient to condense the steam in the cylinder.

Now while the Fire Engine was much superior to the Savery pump in as much as it could pump out deeper mines while using only the flimsiest of boilers working at atmospheric pressure, it was still very heavy on fuel, as the working cylinder was also the condenser. The cylinder had to be brought up to working heat from cold at every working stroke and the obvious solution was to condense the steam in a separate chamber without cooling the main cylinder. Quite a simple matter when viewed in retrospect, but it took the clear thinking of James Watt to design the steam-condensing engine, introduced and patented in 1765.

Meanwhile, although Newcomen had died in 1729, his pumping engines continued to be built and about 1770 a form of condenser was incorporated into the design: not the real thing as we know it—this was covered by Watt's patent and could not be used—but a simple device called the picklepot, a form of jet condenser fitted below the working cylinder.

This picklepot was simply an auxiliary chamber opening directly into the main steam cylinder and draining into a non-return water trap with outlet to the launder. On one side of this chamber was the snifting valve, on the other the condensing spray cock. When steam was admitted to the working cylinder to allow the piston to rise, the air was expelled through the snifting valve and the water through the drain trap. No air pump was required, and if a perfect vacuum was not attainable, at least it evaded Watt's patent and gave a healthy reduction in fuel consumption. This picklepot was used until the expiry of the patent in 1800, when the common jet condenser was fitted.

By the 1780's, the Newcomen engine was made rotative, although still single-acting, and in the early nineteenth century, while it was occasionally used as a colliery winding engine, it was never very successful in that capacity. One of the largest of Newcomen's pumps had a cylinder of 82 inches bore by 11 feet long weighing 6½ tons, and it had three condensing jets. It pumped water from a 570-foot-deep mine in three stages and required three boilers to work it, with another boiler on stand by.

The Newcomen pump was the only steam pump until about 1775–80, and it held its own until well into the nineteenth century. One veteran, built about 1795, was in continuous use until 1923 and was kept in working order until 1931, covering a century and a quarter of useful working life. Burning as it did unsaleable coal from the colliery and costing never a penny in repairs, it owed nothing to the owners when it came to be broken up. Nowadays, of course, we would replace this old pump by an expensive diesel-driven pump unit, burning imported fuel and costing a fortune in maintenance and spares.

Although the cumbersome system of heavy beam and chains gave a very small power/weight ratio, the principle of the Newcomen engine allowed no alternative. The pump rod required a straight pull up and the engine piston gave a straight pull down. Due to the necessity for keeping a deep water seal above the piston, the cylinder could not be placed other than in the vertical position with the open end upwards.

James Watt has been variously described as the inventor of the steam engine, the father of the rotary engine and the discoverer of the elastic properties of steam. He was, in fact, none of these. There were quite a number of Newcomen engines at work before Watt was born, and by the time he was of an age to study the steam engine, about 1760, Newcomen's pumps were clanking and grunting all over the industrial world. Savery had demonstrated the elastic properties of steam some sixty years before that, and rotary engines in the form of windmills and water-wheels had been in use for centuries. James Watt, however, created something even more important to civilisation than the steam engine: the professional engineer. Hitherto, and for many years still to come, the pumping engines had been designed and put together by millwrights and blacksmiths, men accustomed to working in wood or iron by rule of thumb, but from Watt's age onwards, engines were to be built by men trained in the casting and shaping of metals to economical and pleasing proportions, and design was placed in the hands of educated men with enquiring minds.

Watt studied Newcomen's engine and recognised the causes of the high fuel consumption, as others had done before him; then he found the cure in the separate condenser, air pump and lagged cylinder, leaving all other engineers behind him. He also reconsidered the action of steam in a cylinder and found that it could be used to push a piston as well as create a vacuum. Hence Watt's early pumping engine was a true steam engine as opposed to Newcomen's atmospheric engine. Although it used steam on both sides of the piston, surprisingly it was still single-acting: the steam above the piston took the place of the atmospheric pressure. This development was logical, of course, as the mine pumps of the day were still only single-acting, and their long heavy pump rods needed power on the upstroke only, gravity taking care of the delivery stroke.

The vertical steam cylinder was retained but was now fitted with a cover and stuffing box for the piston rod, and steam admission to the top and bottom of the cylinder was by sophisticated cataract gear and drop valves. The steam was condensed in a separate jet condenser with an air pump operated by the beam to pump out the condensate. In action, steam was always open to the top of the cylinder, and with the cold air and water seal excluded from above the piston and the cold water removed from below the piston; with, moreover, the cylinder walls heavily lagged and kept at an even temperature throughout the cycle, Watt's pumping engine had now become a most economical machine. By the end of the 1700's, the wooden arch-ended beam had been replaced by a conventional type of cast iron beam with parallel motion to both piston rod and pump rod, and gradually, by a process of refinement through several generations, this pumping engine developed into the Cornish engine, one of the most efficient machines of the early nineteenth century.

The operation of the piston was controlled by three valves: the steam admission, the equilibrium and the exhaust valves, working in a valve chest giving access to top and bottom of the cylinder, with the equilibrium placed between the two ports. On the downstroke the steam valve was opened to the top of the cylinder, the equilibrium valve was closed and the exhaust valve opened, allowing steam to act above the piston and vacuum below it. At cut off, the steam valve closed and the piston continued until bottom dead point, when the exhaust valve closed

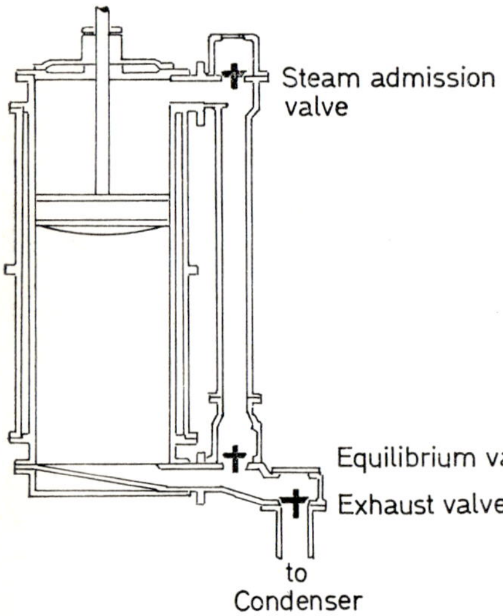

3 *Left:* James Watt's pumping engine cylinder
4 *Right:* SS *Comet* (1812)—Grasshopper engine

and the equilibrium valve opened, putting both ends of the cylinder in communication with each other and allowing the piston to be pulled up by the weight of the pump rods on the beam.

The valve actuating mechanism was controlled by the cataract, a much more sophisticated piece of apparatus than the old tipping bucket, and a most reliable type of stroke-frequency governor. The valve gear was mounted on a loose pump plunger by the beam, but whose descent was retarded by a controllable discharge cock on the pump, releasing the catches on the valves in their proper sequence, and allowing the cut off to be altered at will. This allowed a slight pause at the end of each down stroke to allow the main pump chamber to fill. As with the Newcomen engine, the Cornish engine was merely a pump handle to lift the enormous weight of the pump rods up to 100 tons in a deep mine, and its action depended on almost perfect vacuum. Initially, in Watt's days, the boiler pressure rarely exceeded 10 lb, but in the Cornish engines' heyday, pressures of up to 40 lb were recorded.

As already stated, vacuum was all important, and the engineers of these days had little to learn about keeping the air out of their engines. The piston-rod stuffing box was exceptionally long, with double packings separated by a lantern gland, primed either with steam or with cylinder oil to exclude the air. The air pump and jet condenser were drowned in the feed water tank, again to exclude any air from spoiling the vacuum.

The weight of the pumping gear was compensated, in deep mines, by tumbling boxes or counterpoises on the engine side of the beam, and the inertia of all this massive gear made for a very steady running engine and a smooth-flowing pump.

The economy, and therefore the efficiency, of the Cornish engine was calculated in terms of 'duty', i.e. the number of foot pounds of work done per bushel or per hundredweight of coal, and in the Cornish mines this varied from 18,000,000 in 1813 to 68,000,000 in 1844. One mine engine claimed a duty of 125,000,000 in 1835, equivalent to $1\frac{3}{4}$ lb coal/HP/hour or, for that matter, running very close to the triple expansion engine of the 1890's. It is impossible to check these figures, of course, and one is apt to wonder if they were supplied by the engine-builder rather than the mine-owner. However, they do show that the Cornish engine was a very efficient engine indeed.

This Cornish pumping engine was used almost exclusively for pumping out mines and deep workings, but in some cases it also acted as an elementary form of escalator to convey the miners to and from the bottom of the pit. In the Man Engine the long pump rod reaching down to the lowest part of the mine, with a stroke of from between 10 to 12 feet, was equipped with small steps pitched one stroke apart, there being hand grips at convenient height above each step. The shaft walls were provided with a similar number of platforms of equal pitch such that at top or bottom of the pump-rod stroke, each step registered with a platform. In descending into the pit, the miner stepped from the highest platform on to the appropriate pump-rod step while the beam was at the top of its stroke, and was lowered to the platform next below, getting off at the momentary halt at bottom stroke and waiting a few moments until the pump rod again reached the top of its stroke, when he repeated the process, step by step.

This rudimentary elevator simply bristled with potential risks to life and limb but was accepted as an occupational hazard, the alternative involving an hour's climb up nearly vertical ladders after a hard day down below.

Inevitably a man of Watt's genius soon discovered that an engine capable of pumping out a mine could also be used to provide power for other industrial purposes, and in 1781 he built his rotative beam engine, first of all as a single-acting unit, then in 1783 with a double-acting cylinder, and it is at this stage in his career that poetic justice caught up with him. Just as Newcomen had to make do with a picklepot condenser because Watt held the patent rights for the orthodox jet condenser, so Watt could not use the crank because Pickard was sitting on the patent, but wouldn't use it. Watt was, in consequence, forced to design an alternative method of converting reciprocating motion into rotary motion. This was the sun-and-planet gear, with a fixed cog wheel attached to the end of the connecting rod meshing with a similar wheel on the engine shaft. With wheels of equal diameter, the shaft would make two revolutions to each up-and-down stroke of the piston, and with the slow, cumbersome engines of the day, this method was the equivalent of a 1 to 2 gearing. Upon the expiry of the patent the crank came into its own and from then on was used in all Watt's rotary engines.

These early rotary engines were still being built by millwrights and blacksmiths and were exceedingly crude. One of them still on view, built in 1784, had a square cast-iron shaft turned in way of the journals, a cast-iron connecting rod and a flywheel with a square hole in the hub 2 inches larger than the shaft and packed with wedges. Steam, at very low pressure, was admitted for full stroke, and when started up the engine could just as easily run backwards as forwards; when the valve gear was worked from the beam there was no 'right' direction of rotation.

This, then, was the industrial steam engine at the turn of the eighteenth century, a shapeless heap of clumsy castings with a lofty beam and massive flywheel, turning at about twenty revolutions per minute and housed in a building the size of a cathedral. When man tired of the vagaries of the wind to drive his ships and began casting about for some alternative, controllable power his only choice was a Watt's type rotary engine with all its appurtenances, plus a brick-mounted boiler and smoke-stack for good measure. He discovered, as so many have discovered, that what was eminently suitable for a land installation could be quite out of place on board ship: the sheer weight of the engine would send it straight through the bottom if the high centre of gravity did not capsize the vessel. And the amount of power delivered by this iron monster would hardly cause a ripple in the water alongside.

What was needed was to adopt the central idea of steam driving a piston inside a cylinder, but to rethink the engine with a more direct drive and in a more compact shape, suitable for installation in a ship and still leave room for cargo and passengers. In short, re-invent the steam engine and boiler in reversible form.

This problem was solved in a couple of generations, with inventors coming forward with hundreds of designs for propelling ships based on the expansive properties of steam, and with an embarrassing profusion of crank and valve motions, most of which will be described in succeeding chapters. But it took time for these various engines to be adopted; it cost a lot of money to build a ship and fit a demonstration engine therein, and inventors as a breed are a pretty impecunious lot. Most if not all of the experimental steamers in the early years of the nineteenth century were privately financed, and there was no capital at all for research or development.

If the engine faltered on trials as a result of teething troubles, such as occur on practically every ship ever launched, the vessel risked being left on the sponsor's hands. Considering the number of obstacles in the path of the early marine engines: Britain exhausted by twenty years of war with the French, economic instability, primitive machine production techniques, but principally hostility from the seafaring profession, apathy from the ship-owners and, greatest of all, lack of bunkering facilities, it is a wonder that the marine engine ever came into being at all.

The ideas advanced by this host of inventors ranges over a much vaster field than could be accommodated by the existing technology: they were tried out and failed, not because they were bad, simply because they were in advance of their age. Thus we read of high-pressure engines and water-tube boilers in use among the earliest steamers, which inevitably failed through using sea-water feed. Or of the orthodox crank-crosshead-connecting-rod engine (the Gorgon type), the design for which lapsed and was revived at a later age when the screw propeller was adopted. Many of these designs were experimented with, were found wanting for one reason or other, and disappeared into a smoky limbo for a period, to be re-invented and proved successful at a more advanced stage of engine development or when more suitable materials were obtainable.

By that time, the original steam-engine design had split into three different types: the land engine, the locomotive and the marine engine. The land engine was built as a beam engine for many years and was popular throughout the whole of the nineteenth century, mainly as a rotative pumping engine. It remained as a prime mover well into the twentieth century, and there are several of them in the Midlands still clanking away at 40 lb to the square inch, earning their keep. After a while, towards the end of the century, the beam engine changed to the horizontal mill engine and remained as a popular industrial prime mover until the advent of electricity. But throughout its whole career the land engine remained a condensing engine.

The locomotive engine, on the other hand, could not use condensing plant and the heavy large-bore cylinders suitable for the low pressures then in use. Right from the beginning the railway engine used high-pressure boilers, up to 50 lb per square inch as opposed to the land boiler's 5 lb, and after a few spectacular designs speedily settled down to the conventional, accepted locomotive engine used throughout the nineteenth and twentieth centuries until the diesel-electric engine took over.

The marine steam engine never really settled down at all. Starting as a side-lever engine driving a paddle shaft, it branched out into a multiplicity of designs before becoming a screw engine, first horizontal then vertical, and laboriously working its way up through the expansion range from single to quadruple. Arriving at a reasonable pitch of economic acceptability, it changed in the space of a few years into a rotary engine until challenged by the internal combustion engine.

2 The earliest steamships

From such a long distance in time it is no easy matter to recover details and reconstruct in one's mind the shape of the machinery used in the earliest steamboats. In most cases all records are lost or at best only fragmentary data are available—these vessels came before the days of photography and process blocks—and we are constrained to fall back on contemporary paintings and woodcuts. Even then we are not in any happier position, as the nautical artists of the time, while remarkably good at depicting the beauty of a ship's lines, the trim of her sails or the finer points of her rigging, i.e. a subject they knew and understood, were woefully vague and woolly when it came to describing paddle-wheels and smoke-stacks, or indeed any minor detail of machinery protruding above the deck line.

After all, how did journalists of the 1950's describe the first nuclear-powered steamer to their lay readers?

Since early times various means of mechanical propulsion had been tried in ships in an attempt to put more power into the paddles than was possible by simple rowing: teams of men or horses harnessed to a capstan which in turn drove either a paddle-wheel, an arrangement of vertical oars or an endless belt equipped with paddle boards. All these systems worked just as long as fresh men or horses were procurable and stopped once the muscular power became exhausted. What was wanted was something that would work the paddles as long as they were needed, independently of manpower, and steam offered the only solution. Even so, it was realised at a very early stage in mechanical propulsion that while one man standing on a jetty could push a laden boat out into the stream at a fairly high speed, braced as he was on an immovable base, it took several men sitting inside the boat to propel it at anything like the same speed.

As engineers, therefore, we can exclude from this survey all those experimental craft using wheels, oars or boards worked by muscular power and confine ourselves to a brief description of vessels propelled by different types of heat engines and remember that although the steam engine was invented and developed in Britain, it was the Americans who pioneered the steamboat. While we were building our first commercial steamer on the Clyde in 1812 there were half a dozen steamers already in service round the east coast of North America, and a steam coastal defence vessel was being made ready for launching.

The earliest successful attempt at propelling a boat by steam power was that made by John Fitch, who in 1787 sailed a skiff up the Delaware River by means of twelve vertical oars operated by a horizontal condensing engine using a 12-inch-bore cylinder with a 3-foot stroke. The piston was double-acting and the inventor claimed that a force of half a ton was exerted against each bank of six oars at a stroke.

The following year he built a larger boat, propelled by, apparently, the same engine, and conveyed the crew, plus thirty paying passengers, from Philadelphia to Burlington at a speed of 5 knots. This vessel was followed in 1790 by a stern-wheeler driven by a beam engine and put into service on the Delaware as a passenger and freight ferry. Fitch thus introduced the first-ever commercial steamboat.

Another early steamboat pioneer was the Scot Patrick Miller, who very sensibly used a double-hulled vessel to experiment with, the one to carry the boiler, the other to take the engine, with the paddle-wheels between the two. The engine, supplied by William Symington in 1788, was a form of atmospheric engine with two cylinders whose dimensions are given as 4 inches bore by 18 inches stroke. As the displacement registered some 5 tons and the inventor claimed to have made a speed of 5 miles an hour on Dalswinton Lake, one can only assume that the cylinder sizes quoted for this engine are incorrect. The reciprocating motion of the pistons was converted into rotary motion by means of ratchets and chains driving on to the paddle shafts. This ship was a modest success in a way, but the engine and boiler, to say nothing of the fuel, took up far too much space and left little room for passengers.

Yet another inventor, James Rumsey, broke away from the accepted line of thought on steam propulsion and put a very old idea into a new setting on board: a pump drawing in water at the bows and discharging it at the stern, the old hydraulic jet-propulsion principle which appears from time to time throughout marine engineering history. Rumsey's boat, tried out on the Potomac in 1787, is stated to have made 4 miles an hour, a very optimistic claim. The pump was similar in principle to a modern vertical simplex water pump, with the steam cylinder single-acting, being open to the atmosphere at the top, and using a vacuum for the downward working stroke, as the boiler pressure was barely above 1 lb gauge. This boat, like the others, failed to attract the ship-owners' interest and the inventor's death put a stop to any further development of this principle for the time being.

One of the most famous of the earliest steamboats was the *Charlotte Dundas*, built in 1801 at Grangemouth Dockyard and engined by William Symington, who had collaborated with Patrick Miller and who consequently knew something of the problems associated with marine propulsion. The vessel was built to the order of Lord Dundas, Governor of the Forth and Clyde Canal, as an experimental tug for use on the canal, and was some 56 feet in length by 18 feet beam. The engine was a horizontal direct-acting condensing engine, with a steam cylinder 22 inches bore by 4 feet stroke, a most unusual engine for those days of beams and indirect drive, and which had to be re-invented some thirty years later. The tug was tried out in 1802 and was a huge technical success but, as is well known, a complete commercial failure inasmuch as the canal-owners maintained that her wash would erode the banks. Viewed in the perspective of history, their decision was perfectly justified: it would have cost infinitely more to shore up the canal banks than to hire an occasional cart horse to tow the lighters.

Famous as she was, the *Charlotte Dundas* was only a milestone on that long hard road of marine engineering where success spells a competence during one's lifetime and failure means bankruptcy and ignominy. The experimental tug was laid up in a backwater and allowed to moulder away.

Back across the Atlantic, Colonel John Stevens, of Hoboken, NJ, had constructed a vessel propelled by a pair of propellers driven by a double-acting non-condensing engine with a cylinder 10 inches bore by 2 feet stroke and a boiler pressure of 50 lb. This engine was worked by side rods on to two crankshafts, and steam distribution was by a pair of two-way cocks moved by rack and sector, one at each end of the cylinder. The boiler was unusual for the times and reflected the American practice of high-pressure steam and non-condensing engines, a practice which was maintained in their river vessels throughout the nineteenth century. It consisted of a set of thimble tubes of copper expanded into a steam drum.

The vessel was tried out in 1804 and a speed of 4 miles an hour was recorded by the inventor.

Still on the same side of the Atlantic another inventor, Robert Fulton, had been studying means of applying the steam engine to propel ships, and in 1803 we find him in Paris demonstrating his experimental boat driven by a 'fire-engine' before a large and appreciative audience. According to the Press, the vessel made three miles an hour against the current, and considerably more with it. The engine was an early form of steeple drive: a vertical cylinder with an overhead crosshead working in vertical guides towering above the deck. From this crosshead depended two return connecting rods laying on to a bellcrank lever which in turn drove the main paddle crankshaft through a horizontal connecting rod. This paddle shaft was in turn geared up to a parallel layshaft set lower in the boat and which carried two heavy flywheels to provide momentum and help the crank over the centres.

Steam was provided by a type of water-tube boiler set in masonry working at very low pressure, and apparently some form of condenser was used. The side paddles were simple radial spokes made of wood and fitted with flat boards; the paddle-wheel had still to come.

Naturally enough, this mass of ironmongery, together with the wooden framework supporting the steeple guides, had an enormous weight. The first vessel to be equipped with Fulton's engine, a flat-bottomed lighter, was given no opportunity to try it out: the engine simply fell through the bottom, and a second, heavier boat had to be supplied for the purpose of the trials.

In spite of this success, the French authorities failed to recognise the value of steam propulsion for either war or peace, and Fulton returned to America to pursue his experiments, which culminated in the *Clermont*, the first vessel to be put on a regular commercial run. This vessel was built in 1807 on the East River, New York, and was 133 feet in length by 13 feet breadth, with a shallow draught of only 2 feet 6 inches. Her engine, built in England by Boulton, Watt and Co., was similar to that tested on the Seine, with a cylinder 24 inches bore by 4 feet stroke and developing 19 HP. The side paddles, again only a set of radial spokes, had no outer bearings and were not, originally, encased in paddle-boxes. After initial trials the *Clermont* sailed from New York to Clermont and thence to Albany, the total voyage being made at an average speed of 5 miles per hour. During the summer of 1807 she made regular trips between New York and Albany, and by the end of the year, during the 1807–8 winter, she was lengthened to 166 feet and fitted with cabins, while her paddles were boxed in. She continued year after year on her regular run until 1813 when her name was changed to *North River* and she was put on other duties, but her future had been assured right from the start.

Quite naturally her success aroused a lot of bad feeling among the longshoremen of the district, and her owners had to meet many claims for collision with sailing craft, some of which suspiciously appeared to have been contrived, as well as considerable maintenance to her paddles.

As happens so often in the case of a major invention, another steamboat made its appearance only a few days after Fulton's *Clermont* had been put into the New York–Albany service. John Stevens, seeing no future in his research into screw propulsion, switched his attention to low-pressure steam and paddles, and in 1807 brought out his *Phoenix*, achieving with it a result comparable to that of the *Clermont*. Fulton, however, had won a monopoly for steam traffic on the Hudson leaving no room for any competitor. Undeterred, John Stevens' son, Robert L. Stevens, a name to become famous in American marine engineering, sailed the *Phoenix* from New York round to Philadelphia in 1809, and thus achieved the distinction of making the first deep-sea steam navigation of all time.

Fulton and his patron, Robert R. Livingston, followed up the *Clermont*'s success with other steamers of greater tonnage, most of them fitted with Boulton and Watt engines, and by 1811 he had prevailed upon the American Navy to commission one of his steamers for coastal defence.

Robert Fulton died in 1815 at the early age of fifty, a comparatively poor man, most of his substance having been spent on expensive lawsuits brought about by opponents of steam. His first commercial success, the *Clermont*, constitutes one of the milestones in steam engineering, and was followed in 1812 by yet another milestone: the *Comet*, probably the most famous of all early steamboats, and the first commercial success in Britain.

The *Comet* was built at Port Glasgow by John Wood & Co. to the order of Henry Bell; she was engined by John Robertson of Glasgow and the boiler was supplied by David Napier. She was destined as a passenger ferry to run between Glasgow and Helensburgh, being 51 feet in length with a breadth over paddle-boxes of 15 feet and a loaded draught of 4 feet.

The engine and boiler were placed side by side, the steam pressure being little over atmospheric. The steam cylinder, originally 11½ inches bore but replaced after a brief period of service by one of 12½ inches bore and 16 inches stroke, was placed vertically on top of the iron box frame, the piston rod uppermost. From the unguided crosshead depended two return connecting rods laying on to the ends of two half-side levers, one on each side of the frame, oscillating on a common fulcrum shaft. A crosstail connecting the swinging ends of the half-levers carried a connecting rod which drove the crankshaft placed immediately below the cylinder. This crankshaft carried a large balance wheel and drove the paddle shaft through spur gearing. (See illustration on page 28).

The jet condenser formed part of the engine structure, and the air pump, set inside the condenser, was driven from the side levers by a couple of side rods near the fulcrum. This form of side-lever engine, the half-lever type with the cylinder and crank on the same side of the fulcrum, became very popular for paddle drives and eventually came to be known as the grasshopper engine. It was, in fact, still in use on the Clyde in the 1920's, descendant in a direct line from the original *Comet*. The *Comet*'s funnel was tall and slender, and it doubled as a mast with a yard and sail, a feature common to the earlier steamboats.

The success of the *Comet* was instantaneous and complete: the experimental steamboat had changed overnight into an assured commercial venture and acquired dignity and status. Ship-owners were immediately interested in it and a rash of steamboat builders developed along our main rivers and estuaries. By 1813 a steamer service was inaugurated in rivers as far apart as the Avon at Bristol and the Yare at Yarmouth, while the Clyde could boast at least three steamers registered at Glasgow. By 1815 steamboats were trading to every large port in Britain and the idea had spread to the Continent.

Henry Bell's prototype marine engine had established the framework, the specification upon which most paddle engines were to be designed for the next half-century: low boiler pressure, high vacuum, single vertical steam cylinder driving the paddle shaft through a pair of side levers and reversing by means of a single loose eccentric.

However, before the *Comet*'s design had time to become fixed, before the mould had set, as it were, a large backlog of inventiveness had to be worked out in practice. Henry Bell had created a fashion in steam ferries and before the year was out inventors by the dozen were submitting plans for boilers and engines in a great variety of shapes and sizes, some of them practical but most of them not. Thus several attempts were made to retain the overhead beam, borrowed from the land engine, but in the small wooden vessels of the period the centre of gravity was raised beyond the danger point. Others favoured the bellcrank lever drive, a singularly unlovely and unbalanced device. Some early engines had their cylinders sunk into the boiler, an efficient way of conserving heat but surely a most difficult boiler to design, while others again retained the Newcomen open-topped cylinders. Boilers were of the tank, flue or tube type, rectangular, cylindrical, global or without any geometrical shape at all; at such low pressures shape was not that important. One or two ships even sailed with wooden boilers, like large casks equipped with iron flues and uptakes.

Whatever the type, most early paddle engines worked on the second motion, i.e.the crankshaft was geared either up or down to the paddle shaft through unguarded spur gearing, and in every case in those pioneer ships the engines worked non-expansively. With only a pound or so of steam pressure to play with, steam had to be carried full stroke to get any power at all, and consequently the steam and exhaust pipes were of equal bore.

One of the earliest paddle steamers, built in 1814–15, had a two-cylinder engine worked by overhead beams on to a two-throw crankshaft, the crank angles being at 90°. This crankshaft drove separate port and starboard paddle-shafts through spur gearing, while the jet condenser air pump was driven from one of the beams. The boiler was rectangular and stretched the full width of the vessel, having the two steam cylinders let into it. The engine and boiler thus formed one integral unit. An unusual feature of this boat lay in the method of adjusting the paddle float dip to the draught. Each wheel hub mounted six floats, and these, instead of being fixed to the shrouding, were made to slide inwards or outwards in guides by means of a rack on the inner end of each spoke, the racks engaging in a common pinion movable about the hub. With the engine at rest, this pinion could be eased, jacked round and hardened up again, causing the floats to move nearer or further away from the centre as desired. As the vessel in question was a passenger boat with no cargo space and consequently little change in draught, this refinement seemed rather out of place.

Another vessel of the same period had the engine fitted on the port side and the boiler on the starboard side, the engine being a single-cylinder side-lever unit driving the paddles through gearing in the usual way. The large boiler was of the haystack type and caused the owners quite a lot of worry in that it was almost impossible to keep the vessel on a level keel owing to the excessive changes in boiler level. To a modern engineer this might seem an odd statement in view of the very slight difference in working water-level in a well-managed boiler. But in those far-off days there were no water-level gauges, no pressure gauges, or, for that matter, no marine engineers. The boiler was simply a big tank that one filled up at the start of a voyage and, when the steam gave out and the boiler ran dry, one filled up again. Several commentators of the early days of steam noted that furnace crowns ran at a dull red heat for the whole voyage, but with steam very little above atmospheric pressure there was little danger of an explosion. The greatest danger was that of losing all the steam and causing a vacuum to form in its place: the flimsy boiler shell would simply crumple up like a paper bag.

An interesting breakdown of the cost of building a steamboat shows the very high value placed on the machinery in those days (1816).

	£
Cost of hull	700
Steam engine and boiler	700
Paddles	300
Joinery	200
Upholstery	70
	£1970

It would be very difficult indeed to describe the many early types of steamboat that appeared in the first thirty years of the nineteenth century and the simplest method of studying the development of the marine engine is to take a cross-section of steamships in a busy and well-defined district. Although the *Comet* was built on the Clyde and has in consequence shed a certain lustre on all Clyde-trained engineers ever since, whether deserved or no, steamer construction rapidly spread to the Thames, and in early Victorian days London and its dependencies rated as a major shipbuilding centre. After all, the *Great Eastern*, the largest ship in the world during the second half of the century, was built on the Thames. But before undertaking the study of the ships themselves we could do worse than examine the social climate into which they were launched. Until the advent of the railroad in southern England, in 1838, the only way to travel from the capital to any part of the provinces was by mail-coach if one were rich, horse- or bullock-drawn pack-cart if one were poor, or by boat if one were sufficiently adventurous and tough. The rivers and canals were the main highways of the country, especially the rivers, where a hardy breed of longshoremen usually took over the sailing rights and established a monopoly. This was very much so on the Thames, where the Worshipful Company of Watermen had been founded by Henry VIII, although an association of watermen had been in existence for centuries previously. This body of men enjoyed freedom from impressment by the Lords of the Admiralty to service in the Navy, together with sundry other privileges denied the common man, among which was the power to prohibit the transport of passengers or goods on the Thames except by an appointed waterman in his own rowing or sailing boat. By the 1815's, the passenger tilt-boat was the accepted conveyance between Richmond down stream as far as Gravesend. This was an ordinary wherry with a tilt or canvas shelter at the after end for passengers, who were accommodated with clean straw on occasion to sit on. Rather spartan to our way of thinking, but luxury to the traveller of 150 years ago. For voyages further afield, Southend, say, or Margate, sailing hoys were used, and these could take up to three days, depending upon the weather, while the passengers shared a cabin the size of a telephone kiosk, providing and cooking their own food on board if the rigours of the voyage left them any appetite.

As if that was not enough to daunt the hardiest traveller to the seaside, there were at that time no landing stages or means of boarding the vessels direct except from the quay wall or steps in London. Down river, all passengers disembarked into open boats and thence through the surf to the beach at threepence a head extra, and their luggage was wheeled across the sand in a wheelbarrow.

With the arrival of the steamboat, the watermen saw their trade threatened on all sides, not only in conveying the goods by sailing barge but in the many traffic services across and along the river; and while the general public welcomed the steamers as being more comfortable and weatherproof, speedier, but above all capable of keeping to a reasonable time-table independent of tide and weather, this new form of river traffic met with unreasoning and deadly opposition, which continued for as long as sail lasted.

The earliest steamboats recorded on the Thames were Glasgow built, the first being the *Richmond*, of 50 tons and 10 HP. She had a bellcrank lever-type engine using low-pressure steam from a small flue boiler, and ran between London and Richmond with varying success until her steam pipe burst.

The next was the *Margery*, built in Glasgow in 1813 and brought round to London the following year. She registered 70 tons and was a stern-wheeler with double paddles fitted on brackets projecting beyond the stern, each wheel having radial spokes and duck's-foot paddle boards. Her 14 HP engine was the usual bellcrank lever-type and here again the machinery was not successful, as she spent more time under repair than in service, and when her engine was working her paddles had usually shed an odd board and she had to remain tied up to the bank while a jury duck's foot was being fitted.

She was followed by the *Thames*, built at Port Glasgow in 1813, a side-wheeler with the usual bellcrank arrangement and steam at 2½ lb provided by a wagon boiler.

In 1816 the steamer *Regent* was constructed at Rotherhithe and engined by Maudslay for the London–Margate run. This was quite a large steamer for the times, and she had the by now accepted bellcrank lever drive, with two cylinders 20 inches bore by 2 feet 8 inches stroke, both bellcranks and paddle shafts being of cast iron. The single-flue boiler was situated between the two cylinders. This ship was destroyed by fire in 1817 but was followed in quick succession on this run by such locally built vessels as the *Sons of Commerce, Favourite, Eclipse*, *Venus*, *Albion*, *Dart* and *Magnet*, ranging from 160 to 300 tons and from 40 to 140 HP, their appearance covering a span between 1818 and 1826.

The *Sons of Commerce*, built at Blackfriars in 1817, had a single grasshopper engine (in this context one engine means one cylinder) of 27 inches bore by 2 feet 6 inches stroke, and giving a speed of 10 miles per hour on 20 HP.

One of the most famous of the early London ferries was the *London Engineer*, built at Rotherhithe in 1818 and engined by Maudslay. She was peculiar in that her two paddle-wheels ran in a well cut into the centre line of the vessel, trunked and decked in to form a watertight paddle race open at the bottom. Her engine had two cylinders, each driving through a bellcrank on to one of the paddle cranks.

The wheels were 12 feet 6 inches in diameter turning at 28 rpm and steam was supplied by three single-furnaced copper boilers working at 5 lb. An unusual feature of the machinery was an auxiliary air pump or compressor delivering into the paddle trunk to keep the water-level down and prevent the wheels from choking. By 1822 there were twenty-five steamers registered on the stations between Richmond and Ramsgate, ranging from 45 tons and 6 HP to 315 tons and 100 HP, and a complete network of steam navigation had become established in the estuary in the short space of nine years. Even the earliest vessels were sumptuously fitted out for the times, and compared most favourably with the

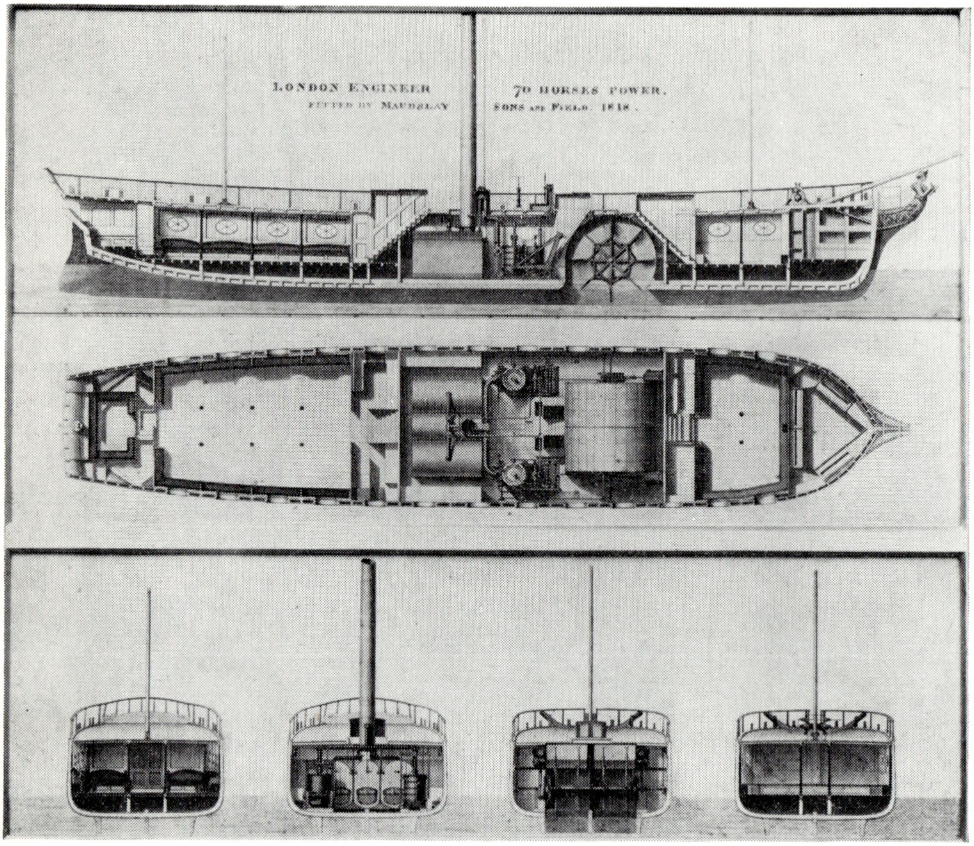

5 SS *London Engineer* (1818)—Bellcrank side-lever engine with paddles in central well

existing sailing hoys. The *Thames* (1813), for instance, had her main cabin heated by a steam pipe running under the seats, carried a stewardess and boasted a select library, though the fare from London to Margate was fifteen shillings per head, rather a lot for those days.

All these early vessels had the long thin wrought-iron funnel serving also as a mast, and were rigged to carry a square sail and several stay sails.

In 1820 river steamers of the *London* and *Diana* type were appearing: 85 feet long by 14 feet beam, with a single-cylinder grasshopper engine of 14 HP, with friction clutches on the paddle shaft allowing the wheels to be worked singly or together and with a hinged funnel for lowering when passing below a bridge.

Another type of engine appearing on the river boats was Joseph Maudslay's so-called vibrating engine, the oscillating engine, first encountered on the Thames in the *Endeavour* in 1828. This had cylinders 20 inches in diameter with a 2-foot stroke at 32 rpm taking steam from a low-pressure flue boiler.

Yet another engine type, of completely different character, made its début on the Thames in 1834 in a blaze of publicity. One of the river companies offered a contract for a ship to be built 'that would, on trials, beat every other vessel of her class on the river and burn less fuel, or be left on the builders hands'. David Napier accepted the challenge and satisfied the outrageous conditions with his steeple engine. This had a vertical cylinder placed immediately below the crankshaft, the piston having two or four piston rods, placed equally on either side of the shaft, driving a crosshead guided by rollers in a steeple framework high above the deck-level. From this crosshead depended a return connecting rod on to the crank.

The vessel was built of iron and had a double-bottom tank in the engine room which served as a jet condenser cum feed tank.

By 1842 the Thames steam navigation had developed into a fleet of sophisticated passenger steamers with locomotive-type boilers working at up to 40 lb pressure and averaging 10 knots. Some of them, the crack ships, could make 18 knots at a pinch—and this in the 1840's! By now most of the freak engines had been weeded out and the orthodox types had become the side-lever, the oscillating and the steeple engines, driving paddle-wheels with feathering floats which, though patented as long ago as 1812, never really became popular until the 1830's. These later vessels were exceptionally staunch and their engines were well designed and built; some of them, re-boilered, were still in evidence by the turn of the nineteenth century.

Although the marine steam engine had acquired respectability and had crystallised into two or three types with minor sub-groups, there were still a few ships trading on the Thames in the 1840's driven by outmoded or outlandish engines that defied classification and that by their very uniqueness tended to be avoided by engineers: the passenger vessel *Cricket*, for example. This steamer, built about 1846, was driven by a four-cylinder compound oscillating engine on to two cranks of the paddle shaft, there being an HP and an LP cylinder oscillating side by side on common trunnions engaging the same extended crankpin. This was, incidentally, one of the earliest compound expansion engines, a type which became popular only some twenty-five years later with the advent of higher steam pressures. The *Cricket* was a very fast boat and extremely economical for her time, making 14 knots on $4\frac{1}{2}$ lb coal/IHP/hour when good average practice could not get the consumption below 5 lb. She had two tubular boilers steaming at 20 lb pressure and these could never be kept steamtight, there being one or other often out of commission for repairs. This vessel's career ended abruptly in 1847 when one of the boilers exploded, killing seventeen people and injuring over sixty.

Another vessel with a peculiar engine was the *Little Western*, a composite ship built in 1841 for the General Steam Navigation Company, equipped with Morgan's patent feathering floats. The engine, built by Acraman & Morgan, of Bristol, was a form of overhead beam engine stood on its head: the single cylinder was placed horizontally on the keelson with the beam oscillating vertically on a fulcrum fitted on to a post between keel and deck. The drive from the beam end to the paddle crank was by horizontal connecting rod. In one way this was an easy engine to manœuvre as this single cylinder had no dead centre. The crankpin with the weight of the connecting rod would normally sink to the lowest point of the crank circle with the engine at rest and this would place the piston at half-stroke, just right for a start ahead or astern. It was not so successful from a maintenance point of view, however. The beam fulcrum bearing, always a most difficult bearing to keep cool, whether overhead beam or side-lever type, gave trouble not only as a bearing but in its anchorage to the ship's structure. This fulcrum takes the combined piston thrust and crank reaction and its anchorage must be on a very solid foundation: side keelsons or engine entablature, but not on a post stepped between keelson and deck beam.

Again, with a horizontal cylinder, the weight of piston, rod and parallel motion gave rise to lubrication and wear problems not encountered in the vertical piston engines.

Yet another unusual engine, that of the *Sapphire*, built in 1842 for the Diamond Company's Thames ferries, had an engine that in principle harked back to the eighteenth-century atmospheric type. There were three open-topped cylinders, set vertically below the paddle shaft, in each of which worked a single-acting trunk piston whose connecting rod lay on to a separate crank, the three cranks being at 120°. These cylinders were 47 inches bore and 3 feet stroke, and the pistons took steam at 8 lb on the outward stroke, the return stroke being by atmospheric pressure from outside acting against a vacuum. The engine, of 100 HP, ran at 34 rpm and the *Sapphire* could make up to 14 knots on the Blackwall–Gravesend run. This three-cylinder trunk piston arrangement's similarity to a modern oil engine was heightened by the use of cams on the paddle shaft to actuate the valves. While this engine was successful enough, it aroused considerable objection on account of its open-top cylinders. In those days when boilers were placed alongside the engine, cleaning fires caused much of the ash to find its way into the gudgeon bearings or on to the piston skirts and score the cylinder walls.

During this period in the development of the Thames and other river steamers the cross-channel services had grown in corresponding measure, although they had been fairly late starters: the original steamboats were essentially river craft, quite unsuitable for open waters. Cross-channel sailing packets had been running to a time-table since 1750 and possibly before, but, unlike the river steamers, their sailing was always conditional on tide and weather.

The cross-channel steamer service was inaugurated in the 1820's in a most unorthodox way when the London to Margate ferry steamer *Majestic*, with 200 passengers on board, was blown out to sea in a gale and drifted into Calais for shelter. The French authorities charged her skipper £24 port dues, but as this amount was not available on board, they placed a couple of Customs officers in charge of the vessel, and prudently removed one of the passengers ashore as security. The last laugh, however, was with the skipper, who broke the deadlock by casting off and sailing speedily for England with the Customs men on board, leaving them to find their way back to France as best they could. History does not record the fate of the shanghaied passenger.

While giving the river steamers a ten-year start, the channel steam packets soon caught up with them and overtook them on the score of sophisticated machinery, and by the 1830's the steam engine had crossed the Atlantic on numerous occasions. The first crossing ever, by the *Savannah*, built in 1818, was purely incidental, and seemed to cause no undue stir in shipping circles. She was a wooden vessel of 320 tons gross, built in New York as a fully rigged ship but converted to auxiliary steam drive after completion, a small engine being bolted to the deck, with portable paddle-wheels. The 90 HP inclined engine, with a cylinder 40 inches bore and a stroke of 5 feet, drove the paddle shaft through the conventional piston rod-crosshead-connecting-rod system direct on to the crank, and the boilers supplied steam at very little above atmospheric pressure, the power exerted being derived from the vacuum in the jet condenser. The side paddle-wheels consisted of radial spokes with floats attached, held in the open position by chains: the paddle spokes could be folded together like a fan when the engine was not required. Under steam alone, and working at 17–18 rpm, the paddles drove the ship at about 4 knots.

The *Savannah* sailed from the port of Savannah for Liverpool in 1819 and made the crossing in twenty-five days under full sail, the engines being used for short periods only when the wind was not favourable, about eighteen days, and she was thus not the first vessel to cross the Atlantic under steam, but the first vessel to use steam in the crossing. Her engine was removed in 1820 when she reverted to a sailing ship, and she was wrecked the following year.

A number of other auxiliary steamers representing various countries followed the *Savannah* in succeeding years: the *Rising Star* (1821), the first steamer to make the western crossing, the French naval vessel *Galibi* (1823), the Dutch *Curacao* (1825), HMS *Rhadamanthus* (1832), and the Canadian topsail schooner *Royal William* (1831), not to be confused with the Irish *Royal William* which was claimed to be the smallest steamer ever to cross the Atlantic under steam in 1837.

These vessels, steam ferries, cross-channel steamers and ocean-going steam packets, were built of timber until the transition from wood to iron took place in the period between 1830 and 1850. Metal had been used for shipbuilding since 1787, when an iron barge appeared on the Severn. The horse-drawn iron lighter *Vulcan* was in use in 1817 on the Forth and Clyde Canal, but it was only in 1820 that an iron steamer was built by and named after her designer *Aaron Manby*. This vessel had one of the first oscillating marine engines, and she sailed from London to Paris to show how an iron steamer could behave at sea.

The master of the *Aaron Manby* had no trouble at all in finding his way across the channel; as a coasting seaman he must have known every headland and rock on both shores by sight, and the phenomenon of compass deviation had not occurred to anyone. But the captain of the iron steamer *Rainbow*, built at Birkenhead in 1837, had a most unpleasant experience on the maiden voyage round to the Thames. Off the Isle of Wight he encountered dense fog and set a compass course that should take him fairly up channel. Hailing a passing fishing boat, he asked for a check bearing and was told he was pointing straight for the beach. With a dead compass and with no landmarks in sight, the captain had to fall back on his years of seamanship and smell his way back to open waters.

Iron ships were unpopular in the early years of the nineteenth century, and it was difficult to find a merchant or underwriter who would trust his goods in a vessel with a fifty-fifty chance of foundering. This was not due to the widely held belief that anything made of iron must inevitably sink when placed in water. Teak is also heavier than water but had been used in shipbuilding for centuries before that. Ship-owners and underwriters were practical men who knew quite a lot about ships, and in their experience no method yet existed of joining together a number of iron plates to make a watertight vessel. Even in the early part of the twentieth century, in the heyday of riveted-steel ships, using smooth rolled plates and hydraulic riveting, a tight joint could not be made without caulking the top plate edge on to the faying plate. In the early nineteenth century iron plates were hammered out, not rolled, and their surface was anything but smooth. Also, rivet holes were drilled by hand with a diamond-point bit and in mating plates these holes, roughly opposite to each, had to be drifted true by force before the rivet could be placed. Plate caulking (in the sense of setting down the plate edge of one plate on to its neighbour after riveting) had not yet been introduced and recourse was had to the other type of caulking, using spunyarn and redlead, or felt, as a joint between the plates, and caulking oakum into the seam after riveting.

This hit-and-miss method of iron shipbuilding disappeared with the advent of higher steam pressures and the necessity for stronger, more steamtight boilers, which in turn demanded better plates and trained men to handle them. The shipwright, the builder of ships, who started off as a carpenter, turned into an iron man about halfway through the 1800's.

Before going on to describe in detail the construction of the early marine engine and the various types of paddle and screw engines which developed during the century, I would like to quote from the log-book of one of the earliest steamers to cross the Atlantic and the first ever to sail into the Arctic, a steamer which, however well authenticated, is not included in the list of early Atlantic crossings. These extracts will give the reader an insight into the many problems which face the marine engineer in any age, and in particular to those which harried the engineers in the early years. It also shows how a strong-willed captain solved all the engineers' problems in one short sentence.

The quotations are taken from the official account by Captain John Ross (later Rear-Admiral Sir John Ross) with the long-winded title of *The Narrative of a Second Voyage in Search of a North-West Passage and of a Residence in the Arctic Regions during the years 1829, 30, 31, 32, 33*, which describes his attempt to discover a route from the Atlantic to the Pacific via the Canadian Arctic. His nephew (Sir) James Clark Ross sailed with him as second-in-command and during that period discovered the Magnetic North Pole. Apart from its engineering interest, the narrative is a stirring account of hardship during four successive years trapped in the ice in the Canadian Arctic, with the loss of only two members of the crew.

The expedition, consisting of one small steamer and a tiny supporting sloop, sailed from London in May 1829, intending to pick up the supply ship, a whaler, in Loch Ryan. The whaling crew, however, refused to sail unless allowed to engage in whaling *en route*, or, alternatively, be paid a certain large sum down in lieu of the catch; conditions which the Captain, with characteristic firmness, refused. Calling in the police to maintain discipline on board the whaler, he transferred all the stores of livestock, food and coal for three years into his own ship and bid the mutinous crew a sailor's farewell.

With his little watchdog escort in tow, Captain Ross crossed the Atlantic under steam when the engines were available, under sail when they broke down, which was most of the time, and reached the west coast of Greenland. Sailing up Davis Strait, he turned west in Lancaster Sound, north of Baffin Land, then down through the Prince Regent Inlet into the Gulf of Boothia, named after Sheriff Booth, the expedition's backer. During the whole of the voyage since leaving London, the two ship's engineers had been sweating down below, alternately patching the boilers and the engines, and it was the engines that finally, in the Captain's opinion, defeated the expedition's aim and lost him the chance of finding the westward passage. His map of the Boothia Peninsula shows it attached to Somerset Island, although in fact there is a navigable channel between the two leading into Viscount Melville Sound and thence to the ice free Beaufort Sea off Alaska. Captain Ross noticed this channel and tried to explore it with a view to finding a way out of the Gulf of Boothia, but an unfavourable wind would not allow intricate navigation through the ice-floes, and his engine was in one of its sulky moods and refused to move. The ship was swept past the inlet, down into the south of the Gulf of Boothia from which there was no return; she remained fast in the ice for three years before being abandoned.

It was a heroic venture, to sail alone into the Arctic with no supply ship to hand, and with an engine that had proved unreliable from the start. Heroic and rather crazy when one remembers that the venture took place a scant ten years after the first-ever crossing of the Atlantic under auxiliary steam, and a good eight years *before* the first Atlantic crossing under full steam.

The log opens with Captain Ross's view on steam navigation in polar waters and for a naval captain of the period his grasp of the fundamentals of steam engineering is extraordinary. He considers steam most advantageous in the Arctic, as the vessel can still proceed in open water when wind is adverse and can explore openings in the ice where it would be dangerous to enter by sail alone. Also a steamer would require less draught and therefore be in a position to avoid floating ice and sail over shoals.

Captain Ross bought his ship in Liverpool, an old Liverpool–Isle of Man packet called the *Victory*, with delivery in London. As the narrative proceeds, it is clear that the shipbuilders had put more faith in the engines than was strictly warranted, and the sails and rigging were in consequence deemed of secondary importance. It is equally obvious that in so far as the machinery was concerned, the brokers 'had seen the gallant Captain coming'. Upon arrival in London, the old engines and boilers were renewed and the paddles replaced by a set of Mr Ericsson's design, being, in the words of the Captain, a patent contrivance that could be hoisted out of the water in one minute when heavy seas or ice threatened them. Her original tonnage was increased from 85 to 150 tons by raising the deck 5 feet 6 inches to accommodate the additional stores, fuel and equipment for a three-year voyage. The engine, from the description given by the Captain, appears to have been of the direct-drive crosshead-connecting-rod crank, and was surface condensing, with two flue boilers working at 45 lb. An unusual feature was the absence of any funnel, the draught being provided by a main bellows worked from the paddle shaft and an auxiliary bellows driven by a donkey engine, and

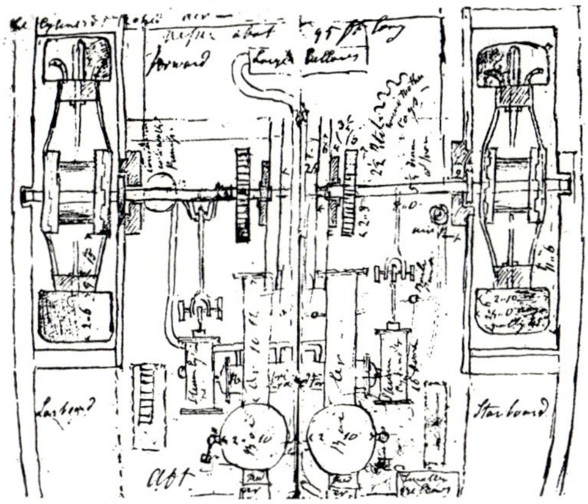

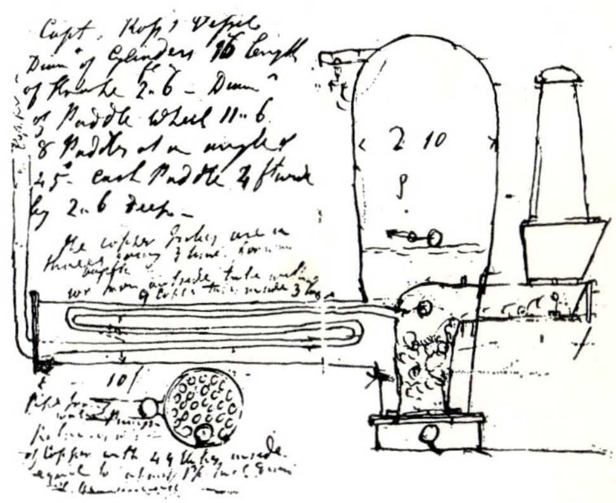

6 *Left:* Plan of machinery of SS *Victory* (1829)
7 *Right:* SS *Victory* (1829)—Ericsson's tubular boiler

with the paddles hoisted on deck there was nothing about the *Victory* to suggest a steamer.

In addition to the normal crew proper to a sailing ship she carried two engineers and three stokers.

The quotations and comments given below are extracted from Captain Ross's log of the voyage, and it would appear that apart from being master of the ship and leader of the expedition he also acted as chief engineer and surgeon.

> *23.5.1829*—'Sailed from London; *Victory* broke down at Woolwich and occasioned delay mending the engine. Proceeded under steam to Gravesend with *Krusenstern* (the escort sloop) in tow. At Gravesend the engine-builders left the ship, and *Victory* sailed on to Margate at $3\frac{1}{2}$-$4\frac{1}{2}$ miles/hour.'

Whilst sailing down the Thames, Captain Ross remarks in an aside that defects in the machinery weigh heavily on his mind. The boilers leaked to such an extent that the hand feed pump was manned to supplement the engine-driven pump. Fresh water was scarce and it was impossible to keep the crew working at the pumps in a temperature of 95°F. Crew exhausted. Not a very auspicious start to a voyage into the Farthest North.

Off North Foreland a favourable breeze sprang up and the paddles were hoisted clear, and shortly afterwards, probably due to the lively seas, the ship was reported to be leaking badly. Whilst under sail, the engineers checked round and found the starboard crosshead guide wheel badly worn. This was removed and brazed up. Also, the crank web keys on the paddle shaft were found to be loose. The Captain was disappointed in this as he felt that the engine manufacturers must have known about such a defect but never disclosed the information and there were no spare keys or key material, i.e. key steel, on board.

> *30.5.1829*—'The principal key on the starboard side broke and starboard paddle out of commission. A new key made of iron but this gave out soon after. Three new iron keys were made and fitted in succession but each lasted only an hour before being broken up.'

Rounding Land's End, the boilers continued to leak badly and there was great difficulty in keeping steam. Following the engine-builder's instructions as given before sailing, cow dung and potatoes were shovelled into the boilers in an attempt to make them tight, on the same principle as the use of sawdust to stop a leaking condenser in later years, but all to no avail. This garden compost may have been suitable for low-pressure boilers, but not the *Victory*'s, and they continued to leak so badly that all hands were exhausted not only by pumping up the boilers but by pumping out the bilges.

Finally, the condenser broke down completely with the air pump constantly flooding, and after some discussion, and in view of the severe shortage of fresh water, the engine was converted to jet condensing. Surprisingly, the revolutions increased slightly to 16 rpm.

A day of two later the great bellows, that worked from the main shaft, was disconnected in an effort to save fuel, and the small auxiliary bellows put in service, but it was evident that this latter machine would not last very long. In another aside in the log—the Captain is much given to asides, possibly to relieve his feelings—he complains that everything about this engine is imperfect—the cylinders were too small to start with, and there was little to expect from an engine that could scarcely move the ship at 3 miles per hour. However, he does praise the designers for the absence of funnel, as in the sailing ship–steamer combination, it was always that smoke- and fire-belching funnel which ruined the sails and rotted the rigging and foot ropes.

Once into the Irish Sea, the leaky caulking took up and the crew could unship the hand pumps to the bilges, although the boiler feed still had to be supplemented by hand pump. With all sail on, the *Victory* was found to be a poor sailer.

4.6.1829—'Put into Isle of Man for new key material.'
6.6.1829—'Left Isle of Man, engine speed 15 rpm. Engines broke down.'
7.6.1829—'Off Mull of Galloway. Lee paddle only in water, making 18 rpm, found we could back to windward as well as any vessels in the company, even gaining on them.'

Off the Mull of Galloway, William Hardy, the senior stoker, slipped whilst oiling round and his arm was caught in the guide and crushed by the guide wheel. As the surgeon was not yet on board but was to join the ship in Loch Ryan, Captain Ross amputated the stoker's arm and apparently made an almost professional job of it, thus adding to his already imposing list of qualifications. All speed was made to reach Port Logan, on the Mull of Galloway, to land the patient, but the spur wheel driving the auxiliary bellows gave way with a loud report. Captain Ross rather despondently wrote in his log that he considered that not a single item of this machinery could be aught but a source of vexation, obstruction and evil. Whilst writing this, and to prove his point, the boiler joints gave out, the water poured into the furnaces and put out the fires, thus bringing the *Victory* to a full stop.

9.6.1829—'Reach Port Logan and find that the crew of the *John* (the supply escort ship) had mutinied and refused to sail with the expedition.'

Upon Captain Ross's return to UK in 1833 he learned that this same crew of the *John* had mutinied again in Baffins Bay in 1830. What is astonishing is that the two engineers and the two remaining stokers, having rebuilt the engines half a dozen times in the space of a fortnight and thus having a taste of what was in store for them during the coming voyage, did not quietly pack their bags at Port Logan and take to the heather.

10.6.1829—'Land small boiler and apparatus for cutting ice as these are mere lumber and do not justify the space and weight.'
11.6.1829—'Checked round machinery and found failures in the iron cement in the boiler joints. No spare cement obtainable. Repairs also needed for both main and auxiliary bellows.'
13.6.1829—'Left Loch Ryan leaving *John* behind.'

Captain Ross must have been a truly outstanding leader to command such trust in his crew: to set out on a voyage of exploration into the Arctic lasting several years in a ship barely seaworthy, a bad sailer and with crazy engines was bad enough. But to sail into the unknown without the material and moral support of a supply ship was sheer heroism. The little *Krusenstern* was barely bigger than a lifeboat and could not possibly be considered as a get-you-home unit under any circumstances.

The *Victory* left Scottish waters and sailed north of Ireland, but off Rathlin Island the fore topmast broke.

> *16.6.1829*—'Split jib, found that the storm jib had been stolen in London. Pumps going all the time.'
> *25.6.1829*—'Still repairing engines and boilers, also bellows.'
> *27.6.1829*—'Boilers now tight. Engineers connect feed pump to small engine to save constant labour.'
> *6.7.1829*—'Off Cape Farewell.'
> *11.7.1829*—'Engines out of commission.'
> *13.7.1829*—'Engines tried out but favourable breeze sprang up and they were not required.'
> *16.7.1829*—'Too much swell for engines.'
> *17.7.1829*—'Lee paddle engine working at 17 rpm. After three hours one boiler leaking, put out for repair but the other not sufficient to keep the paddle going. Found port flue pressed flat, edges rent and largest flue flat. Replaced largest flue and repaired port flue. The screw holes in the flanges did not correspond with each other, had to bore new holes after plugging the old ones. Also, the flanges did not meet. Eventually had to renew the port flue.'
> *18.7.1829*—'Engine working lee paddle. After half an hour the main key to the shaft gave way. Port boiler leaking, but screwed up flanges, making new key. Paddle hove up clear of water but tackle gave way and paddle fell onto its bearing. No damage.'

In passing, the use of the word 'port' in the text for the left-hand side cannot be explained, as the common expression was 'larboard' until well into the sixties.

> *20.7.1829*—'65°N by 55°W. Off Greenland. Everybody who could work was stationed in engine room repairing machinery. One new key fitted to shaft.'
> *21.7.1829*—'Engines started but boilers still leaky and only one usable. Speed one mile per hour. Stopped to clean furnaces. Started again.'

22.7.1829—'Small bellows engine useless, has to be worked manually, watch and watch. Engines started on lee paddle at 10 rpm. Engines stopped due to swell and wheels hoisted on board. Engineers took opportunity to continue the never-ending repairs.'

28.7.1829—'Off Disko Island.'

29.7.1829—'Engines started, stopped again as wind fair. Engines started and stopped as required.'

1.8.1829—'One boiler still leaky.'

3.8.1829—'Engines stopped to repair main feed pump.'

4.8.1829—'Pump repaired but starboard boiler still leaking, giving only 10 rpm on one boiler and one paddle wheel, at 1¼ miles per hour. While engines could do 1¼ miles per hour, this was better than nothing, though our debt to the engines was assuredly as small as it could well be. Leak in boiler stopped, engines started. One stoker nearly suffocated by inhaling sulphurous gas at the furnace mouth.'

5.8.1829—'Engines working interruptedly 14 hours, but this day worked 12 hours on end without stopping; longest period yet experienced on this voyage.'

6.8.1829—'Engines behave for 24 hours. 73°N by 74°W. In Lancaster Sound. Feed pump broken again.'

7.8.1829—'Steam at 30 lb only.'

8.8.1829—'Engines being used all the time at 11 rpm. It required constant and minute attention to persuade it to work at all. Even with all the goodwill of the workmen, my presence was for ever required in the engine room, insomuch that I was scarcely allowed to sleep.'

9.8.1829—'Men still at bellows to keep fires going. 74°N by 77°W.

11.8.1829—'Entering Prince Regent Inlet.'

12.8.1829—'Steam up but feed pump broken down.'

13.8.1829—'With defective feed pump engines can make only 8 rpm, or one mile per hour.'

A new method of steam sailing in ice-littered waters was tried out and found successful. A whale line was carried out about two cables' length and the *Victory* warped ahead, during which time a head of steam could be raised on the one boiler. The engines were then set ahead and the ship steamed as far as she could while the pressure lasted (about a quarter of an hour), sending out the warp for another bite and so on throughout the day: alternately steaming and warping along every quarter of an hour.

15.8.1829—'Near magnetic North Pole.'
17.8.1829—'Proceeding down Prince Regent Inlet. Engines used on and off, not behaving well. 71°N by 94°W.'
26.8.1829—'Ship in ice most of week. Off Boothia Felix (Boothia Peninsula). Engines performed badly, 1 mile per hour, at 7 rpm. Anchored off Port Logan (newly discovered) 71°N by 93°W. Boiler pipes being constantly cleared of coal dust.'
31.8.1829—'Still in drift ice.'
13.9.1829—'Still in drift ice.'
15.9.1829—'Sailing all this while either by warping with the boats or fast in the ice. Warping the whole morning to gain four ships' lengths.'
30.9.1829—'Fast in the ice.'
1.10.1829—'Frozen in for the winter.'

Here Captain Ross gave the order that proved beyond any possible doubt his capacity as the leader of an Arctic expedition, an order which in one short sentence resolved all the engineers' and stokers' problems, an order, in fact, which thousands of marine engineers of my generation, struggling with salted up boilers or cantankerous diesels, would have given a year's pay to hear: 'Remove the engines ashore.' True to naval tradition, where nothing is wasted, the engines and boilers were cut up into portable bits, transported ashore on sleds and heaped into a cairn on a low hill. They were not allowed to be merely dumped on to the ice to sink to the bottom in the early summer. The log continues:

3.10.1829—'Men employed taking the engines to pieces to land them.'
12.10.1829—'Engine-room bulkhead removed and seamen's quarters enlarged thereby. Totting up, there are 700 bushels of coal and coke on board, sufficient for 700 days. Part of engine hoisted on to the ice.'
16.10.1829—'Getting out the boilers.'
19.10.1829—'Engine frames brought up.'
20.10.1829—'Last part of the engine brought up on deck.'
22.11.1829—'Engines on the ice, being dismantled by the engineers.'
24.11.1829—'Still cutting out various ironwork of the engine.'
24.4.1830—'Engineers cutting up boilers on the ice to obtain new sheathing for stem where original iron had been torn away by the ice.'
28.7.1830—'Some parts of the engine reshipped that might be convertible to general use on board.'

The *Victory* remained frozen in during the whole of the summer of 1831 and the winter 1831–2 and was abandoned at the end of May 1832. The following winter (1832–3) was spent farther north on Somerset Beach and the expedition was finally rescued on 26th August 1833 by the *Isabella* of Hull.

On a desolate beach off Cape Margaret, in the Gulf of Boothia, under a black Arctic sky and in the eternal silence of the Frozen North, lie the bones of one of the earliest marine engines ever built and the first to venture upon a Polar expedition.

3 Early marine engineering

We have just seen what marine engineering had to offer its devotees in the early years of the nineteenth century, and while working conditions may have improved with the passing of the years, and pay and status changed very much for the better, the marine engineer's calling still demands qualities and endurance denied the ordinary mortal. Professional skill goes without saying; patience beyond reason without a doubt; obstinate determination to complete a job for the well-being of the machinery as a matter of course. But what landlubber could move freely from the superheated atmosphere of a turbine engine room at 140°F to the below-zero temperature of the refrigerating machinery brine room? What sort of man could work on the boiler tops at sea to pack a gland or remake a joint in a temperature of 160 F? What manner of fool would crawl along a plank laid over red-hot ashes in a boiler under steam to expand leaky tubes in the combustion chamber at a temperature well above the boiling point of water? Only a marine engineer, a species of salamander indifferent to either heat or cold, and as much at home in a withering heat among madly whirling machinery as immersed in cold water up to his neck cleaning choked bilge strums.

The *Victory*'s engines certainly gave a lot of trouble to her engineers and firemen, but she was not unique in that respect: early marine engines were in the main still experimental, and anybody who has ever sailed with an experimental engine will confirm that it will behave only for as long as the builders, with their specialised equipment, are on board. Watchkeeping has not changed very much in essence since those early days: although modern propelling machinery is more sophisticated and better built, there is far more of it in the present-day ship than in the early Victorian steamer with its single rickety engine, and it has been very truly said that 'What you don't fit don't give no trouble'.

Working conditions have improved very considerably with the introduction of a three-watch system, electric lighting, mechanical ventilation, well-placed ladders and gratings and, latterly with the development of automation, centralised controls for the main and auxiliary machinery and all the essential services, nicely situated in an air-conditioned and soundproofed cabin. On the other hand, mechanical engineering has increased in scale as well as in efficiency, and taken by and large from personal experience with such varied engines as oscillating engines, geared turbines and diesels, there is little to choose from in matters concerning one's peace of mind on watch. Some simple engines required nursing all the time to prevent piston rods from heating up or eccentric straps from chattering. Others, the double-reduction geared turbines, gave little trouble at sea, but the watchkeepers were forced to put up with the terrible, high-pitched and nerve-shattering scream from the gearing. Others again, the early diesel engines, demanded engineers whose system was proof against the fetid atmosphere of diesel fuel and hot lubricating oil, and could withstand the wear and tear of constant breakdowns at sea with all hands turned to for repairs.

General preference seems to lie with the old-fashioned triple-expansion engine, the matiest and certainly the most reliable of all marine engines and the only type with which one could establish a sort of affinity, such as, for instance, the cowboy and his horse or the shepherd and his dog. Kipling's chief engineer McAndrew thought of his engine as having an individuality of its own, and this concept is shared by many engineers of my generation. But this fancy is not evident with turbines and internal combustion engines.

The earliest steam vessels from the time of their general acceptance until, say, the 1830's, were invariably paddle-driven, for, although the principle of screw propulsion had been discovered many years before, no one had yet worked out the mechanics of a marine propeller or how to drive it with the slow-moving, ponderous engines of the period. These paddle engines were of many types but the most popular of all were the side-lever or the grasshopper engine. The vessels were made of wood until the 1830's when iron plates of suitable size for shipbuilding made their appearance, and the period between 1830 and 1850 may be considered the transition stage between wood and iron ships, after which they were usually constructed of iron.

In these early wooden steamers there was little, if any, collaboration between shipbuilder and engine constructor and this is not surprising. The science of naval architecture, which correlates the geometry of the hull with the mechanics of propulsion, was not yet born and nobody yet had thought of the integration of hull and engine. The ship was built with an empty space roughly amidships and the engineers were expected to instal their engine and its boilers into this space as commodiously as possible. The paddle shaft usually lay across the deck supported on plummer blocks, and in the earliest models the paddle-wheels were simply radial spokes supporting each a paddle board. These soon developed into rimmed paddle-wheels, at first exposed, then boxed in, and eventually, after a number of collisions had shown the necessity, they were protected by sponsons, an extension of the deck beam shelf around the wheels. Some ships came to have an outside paddle bearing on this sponson and so eased a large part of the weight of the wheel from the crankpin.

A typical engine room of that period was a two-tier affair, the upper storey bright and shining, the lower part, the basement, a dark and noisy cavern. The steamer was still a sailing ship with flush decks relieved by the minimum number of housings and skylights. The engine casing was as small as possible consistent with the working space round the crankshaft and running gear, and was surmounted by a thwartships skylight. The crankshaft, at deck-level, was supported on a heavy cast-iron box frame, the entablature, which in turn rested on four cast-iron pillars from the engine bedplate. The running gear, the throttle and reversing controls were all placed at this level, and the upper part of the machinery space, forming the watchkeeping station and manœuvring platform, was the only part that enjoyed any natural light at all. The engine room was completely decked over except in way of the crankshaft entablature, and any sunlight that found its way through the crank races in this box-frame soon lost itself in the general darkness down below.

Beneath the crankshaft level, down on the bottom of the ship, the great iron side levers seesawed up and down in perpetual gloom, their trunnion bearings adding their grunts to the diapason of clicks and clanks from the air, bilge, feed and brine pumps cluttered round the baseplate.

None of this machinery was fenced in or guarded by handrails, while the working platform consisted of oil-sodden planks laid across the floors, some of them with unsupported ends which, if carelessly trodden upon, precipitated the unfortunate attendant into the bilges or, on occasion, below the side levers. In the days before electricity, gas or paraffin, oiling round a marine engine was rather like performing a flying-trapeze act without the safety net.

For economy's sake, only one hurricane lamp would be lit down below, and each watchkeeper carried a duck lamp on his rounds: a sort of continental tin coffee-pot full of colza oil with a spongy wick in the spout emitting a smoky reddish glow.

The early marine engines were very inefficient and this we can understand: the engineers were still groping their way towards higher steam pressures and better boilers. But this does not explain why, when we compare their weight, strength and large bearing surfaces with those of the modern light-weight engines with a hundred times the power, they make such a poor showing. Was it bad design, poor materials, bad workmanship or inefficient personnel? It was, in fact, all of these, but we must be fair to the engine-builders of the day, who not only designed, built, installed and ran them, but also carried out their own research into steam. At that time there was no pool of accumulated experience they could tap, every firm was on its own. The builders were essentially practical men who had been taught engineering the hard way; they thought with their hands, and education, for them, meant only the ability to write their name and count their pay. They were still experimenting with steam propulsion and some of their efforts had to fail if only to achieve a modest average performance.

Materials were of the simplest: cast iron, wrought iron and brass for bearings—steel in any quantity was still a long way in the future. Workmanship, i.e. craftmanship, was as good then as it is today, but of a totally different type, for in the early nineteenth century marine engineering was little more than a cottage industry, a minor and troublesome offshoot of the millwrights' trade where everything was made by hand—machine tools came long after the steam engine, not before it.

The marine engine was born in the Regency period, an age when elegance was the master virtue and design and proportion the key notes. Everything associated with this period—architecture, furniture, carriages, manners, speech, clothing, firearms, clocks, *et al*—reflect a refinement of taste and a polished and competent manufacture. Everything, that is, except the steam engine. There is no good reason why the steam engine alone should show such a lapse in good taste.

The eighteenth-century land engines were shoddy in the extreme and consisted of a rough cast-iron cylinder balanced on a wooden framework driving a wooden beam through an iron chain and controlled by a clutter of forged-iron spindles, levers and catches composing the valve gear: a huge ungainly heap of ironmongery that had neither definite shape nor proportion. Cylinder-bottom flanges had the bolt holes cast in, not bored, and their corresponding holes in the cover seldom coincided. The cylinder itself was simply clamped down on to the foundation beam by dogs and black bolts. Shafts were square forgings filed round in way of the bearings and the bearing blocks themselves were held in position by either dogs or coach screws. The flywheel and sometimes the crank web, both castings, had square cast holes a loose fit on the shaft and locked in place by multiple keys or wedges. Piston rods were plain forgings filed smooth along the working surfaces, and connecting rods were of the spade type, with separate strap heads and cotter adjustment.

All this in an age when the armourer and the silversmith had raised the art of metal working to a pitch undreamed of today.

One possible reason for this inconsistency is that the engine was the product of an uneasy union between millwright and blacksmith. The former, in spite of his knowledge of motive power and mechanisms, was trained to think in terms of wood, he could not build a steam engine without calling in the blacksmith, and in the event it was the blacksmith who took over the building. As it is, we got the working engine, but it took two generations to develop a breed of men who could not only think in terms of iron but who also knew how to work it. However, before they could start to work it they first had to invent power machine tools for the purpose. As the power was supplied by the steam engine in the first place, it took time and a good deal of ingenuity to break this vicious circle.

The earliest marine engines belonged to this period and in general their design was of the poorest. The most common type, and for a while the only type, was that using a pair of side levers to drive the crank. This engine was a complex of discrete components, tied together by the fabric of the ship itself, and was commonly referred to as a house engine. It had a single vertical cylinder bolted direct on to the keelson while the trunnion bearings of the side levers were bolted to the side keelsons. The crankshaft was supported on the deck beam or on brackets built out from the ship-sides, and in a seaway, with each bearing moving in different directions in unison with the working of the ship, the engine would be grunting and labouring to keep the paddle-wheels turning, much of the effort being absorbed in friction. Holding-down bolts for cylinders and seatings were quite often through-bolts driven up through the bottom of the ship and almost impossible to harden up with the ship afloat. Bearings were 'run on the bolts', i.e. adjusted to bind hard on their journals and then slackened off an amount considered reasonable to the engineer. Lubrication was by tallow, both inside and outside the cylinder, supplemented by sea water on the larger bearings.

The second generation of marine engineers introduced the bedplate and its associated entablature to bind together into a rigid framework the various working parts of the engine, keeping the cylinder truly vertical and the side levers working together, while the weight and strain of the paddle shaft was removed from the deck and taken up by the entablature and thence through heavy cast-iron pillaring to the cast-iron bedplate. In this way the thrust of the paddles was absorbed by the bedplate and distributed throughout the framework of the ship's bottom.

The very earliest steam engine used a cylinder cast in yellow metal, but as the size of cylinder increased, economy imposed a cheaper metal, and steam cylinders since those early years have invariably been made of cast iron. Looking backwards from the lofty vantage point of this nuclear age, where every industrial process is automatic, it is very difficult, if not wholly impossible, to appreciate the problems that had to be solved in building a steam engine of any size.

Take, for instance, the steam cylinder. This would arrive from the foundry as a rough, a very rough, green-sand casting that would require several servicings of pickling and hammering to get rid of the vitrified sand. After that it had to be 'skinned', the casting skin had to be either chipped off or worn off by rubbing down with a leaden slab cast to the curvature of the cylinder bore, and smothered with oil and powdered glass. By this slow, laborious process the inside of the cylinder was worn smooth enough to offer a surface to the piston packing or ring. It will be noted that while the surface finish could be improved, the circularity of the bore remained as cast, until such time as a cylinder-boring machine was developed. James Watt himself, that master engine-builder, is claimed to have been eminently satisfied if any one of his engines could boast a cylinder bore circular to within $\frac{3}{8}$ inch.

The earliest boring mills were those built for iron cannon and pump barrels and they worked on the same principle as the machines for boring out trees to form wooden pipes for waterworks. The cast-iron cylinder was chained in a horizontal position on to a heavy trolley running on rails, so that it could be very gradually advanced or fed towards the borer by tackle and hand windlass. The borer consisted of a long boring bar carrying a cast-iron circular head at its end, the diameter of the head being only slightly less than the 'as cast' cylinder bore, and fitted with a number of small steel cutters wedged into its periphery. The borer was driven through wooden gearing by horses or, where available, by water-wheel. In action, the cylinder was fed on to the boring cutter by careful manipulation of the windlass until by degrees a complete traverse had been made and the more prominent lumps and projections had been removed. A second, third and subsequent cuts were then made until a reasonably smooth and consistent surface had been obtained. As the boring head was unsupported, all the weight was taken by the bottom cutters and boring was almost entirely confined to the lowest part of the cylinder. To remedy this, the cylinder had to be turned round 90° and the boring recommenced, four shifts being required for a completed boring out, leaving the cylinder slightly lobed in consequence, although this would have little effect on the engine efficiency at the prevailing low pressures. With the cutting head rotating at one revolution per minute and with an average of three cuts per sector, or twelve cuts in all, it took weeks to bore out a large steam cylinder.

With improvements in foundry technique and better choice of ores, steam cylinders came to be supplied in phosphoric cast iron, a material rich in phosphorous and with a low melting point and consequent increased fluidity, giving sharp impressions and an excellent casting skin. This material was not too strong: with a 2% phosphorous content the tensile strength would not exceed 10 tons to the square inch, but it was soft and possessed very good wearing properties. Above all, it gave the cylinder a very smooth casting skin that required little, if any, machining. An instance of this type of casting is seen in the old-fashioned kitchen range, where the mouldings were burnished to a silver finish from the as-cast condition.

So much for the steam cylinder, but what about the crankshaft and all the wrought-iron articulations—how were they produced? Here again the craftsmen of the period had to solve problems of technique almost impossible for us to imagine. Some engines used cast-iron crankshafts, built up of separate cast-iron journals, with webs keyed on, and with a removable crankpin, but their manufacture, such as it was, was dealt with quite simply by hand tools. As far as wrought shafts were concerned, they were forged 'up' as opposed to our present-day method of forging 'down' from blooms or ingots.

The modern internal-combustion-engine crankshafts can be produced in a number of different ways, depending mainly upon their size. Those for the small petrol and diesel engines may be of spheroidal or nodular cast iron, an extremely strong and tough material. The majority of the small engines, however, say up to about 300 BHP, have their crankshafts formed from a length of heated steel bar shaped in a split mould or die under a drop hammer, the lower half of the mould being fixed to the anvil while the upper half is fitted to the hammer tup. This tup, which can weigh four tons or more, is raised rapidly by a belt over a friction clutch riding an overhead shaft running at constant revolutions. By engaging and releasing this clutch, the tup can be lifted and dropped with the speed and power of a steam hammer, and the two halves of the die will be brought together in a few blows, compressing the steel bar into the shape of the mould.

There are, of course, certain limitations to this method of drop forging, one of them being the necessity for accuracy in cutting the correct lengths of billet from the bar stock, and this is not just a matter of dimensions as much as one of volume. The crankshaft, when forged, must completely fill the die and overflow between the interfaces of the mould. This flash or fin, which is sheared off while still hot, ensures that there will be no void spaces in the forging and also safeguards the costly dies from chipping or cracking by offering a soft buffer between the two mould faces. Another point to bear in mind is that the crankshaft, when forged, must be removeable from its die. This is not really a serious objection as most small diesel engines are of the four-stroke type and in the case of two- and four-cylinder units their crankthrows are at 180° to each other $\left(\frac{720^\circ}{\text{No. of cylinders}}\right)$ With a six-throw shaft the dies cannot be made with horizontal interfaces, these have to be stepped or inclined to the horizontal to allow the crankthrows, at 60° round the circle, to be eased out.

Intermediate-sized crankshafts can be forged direct from blooms or ingots. In this case the crankthrows are slabbed down into roughly rectangular shapes, either to one side or the other of the journal centreline, but all in the same plane. The journals are rounded under the hammer by forming tool, and when the forging is shaped, each throw is twisted round to left or right to give the required crank angle, all the twist being taken up in the adjacent journals, and no twist will be more than 90°. Before machining, the centre part of the crankthrow is cut out to form the webs and pin.

Still larger engines have their pins, journals and crankwebs forged separately then built up by shrinking the webs on to their pins and journals in the required sequence. In some very big shafts each crankthrow, consisting of two webs and the mating pin, are cast in one piece and machined separately before shrinking on to the main journals.

The most modern method, the continuous grain flow (CGF) method, is to progressively mould each crankthrow under heat and pressure into its proper shape and angle in a die and in principle it harks back to the drop forging technique. But in all these different ways of fashioning a crankshaft the mass of steel is taken for granted, sufficient for the job with a bit left over to form the porter bar by which the shaft is manipulated under the hammer or press.

At the beginning of the nineteenth century manufacturing techniques were far more difficult. There were no steam hammers to do the forging, no electric or steam cranes to handle the heavy shafts and no high-speed oil furnaces to heat them. But above all else there were no iron billets or blooms at hand to make them with.

To manufacture a crankshaft in the early days of steam, every little scrap of iron was collected and laid in a corner handy to the coke fire: old bolts, horseshoe nails, scrap iron, boiler plate, angle iron, etc., and these were heated up, a few at a time, and hammered into a homogeneous lump. A few of these lumps were then heated and forged into a slab, the process being repeated as often as required to build up a shaft or a web. During this process of piling the slabs, the material was kept under constant hammering to squeeze out the slag and obtain a high degree of purity in the iron, although this ideal was not always achieved, the best of irons having only 60% of the strength of mild steel. A leading forgemaster of the period claimed that no iron shaft was ever forged that had no flaw in it, and if that flaw happened to be located in a stressed part, the shaft would fail. Shafts did fail, of course, especially those with solid forged cranks, as the discontinuities inherent in a piled material often carried through the thickness of the web.

More generally, however, the shafts were built up of separate forgings, the webs being keyed on to the journals and the crankpin being either bolted or cottered into position between the webs. Shafts and webs were not machined except in way of mating surfaces and the finish was mainly by hand. Crankshafts as often as not were forged square with a rough-turned or filed journal.

Before leaving the subject of the quality of forged iron it might be of interest to compare the chemical analysis of an ancient Greek iron with an early Victorian iron:

	100 BC	AD 1840
C	·085	·045
Si	·004	·032
S	·005	·251
P	·018	·160
M	trace	—
Fe	99·888	99·512
	100	100

The process just described of building up a wrought-iron shaft will explain why connecting rods were kept to the spade type as it would have been an exceedingly difficult and costly job to produce a forked rod—single crossheads were the obvious design.

As steam engines became more and more powerful, requiring heavier crankshafts, piston rods and connecting rods, forgings reached a stage where the smith's hand-wielded hammer was ineffective, and some mechanical type of hammer had to be invented. The only mechanical hammer available, the tilt hammer, remained in general use until the 1840's when the steam forging hammer took over, but the old tilt hammer lingered on in remote country areas well into the twentieth century. It was a useful and handy tool, its only drawback being that by the nature of its construction it could not forge a parallel slab.

It consisted of a wooden beam mounted on a fulcrum with a heavy hammer block at one end and an iron striking plate at the other. A rotating shaft driven either by horses or water-wheel, and carrying a cam or star wheel, was mounted just above the beam striking plate, such that in action the end of the beam was rapidly depressed and released giving that steady rhythmical beat at the hammer head so essential to good forging.

The steam hammer, more or less as we know it today, was invented by Nasmyth for a special purpose, that of forging the 30-inch-diameter paddle shaft proposed for Brunel's *Great Britain*, conceived in 1839 and finally built in 1845. No forgemaster in the country would consider such a gigantic undertaking and the designer was left with the problem of experimenting with a cast-iron shaft and finding somebody who could supply the castings. In the event Brunel changed his mind and built the ship as a screw steamer, but there is a very strong possibility that the difficulties in obtaining forgings of the required size had some bearing on his decision. Nasmyth's design for a steam hammer that would make a good forging of the size required was not taken up by the local forgemasters and had to wait until 1842 before his hammer was adopted.

Originally, steam was used only to lift the piston and the hammer fell by its own weight, i.e. the cylinder was only single-acting. He soon altered the valve arrangements to make the piston double-acting, and so speed up the work, and it is odd that while the steam engine itself has changed so much since Nasmyth introduced his hammer, the steam hammer itself is still very much as he originally designed it.

Nasmyth was a great believer in labour-saving devices, and in modernising plant in factories. He had a very good working knowledge of foundry practice, and one of his ideas was the mechanical hanging skimmer used to retain and hold back the slag in the ladle when pouring a casting. One of his best inventions with possibly as far-reaching consequences as his steam hammer was the safety foundry ladle introduced in 1839. This was the hand-controlled worm and worm wheel to tilt the ladle and ensure a steady pour. In the old method the ladle of molten iron was suspended over the mould and tilted by means of long capstan bars by a team of men: a heavy and very dangerous job.

The arrival of a set of forgings at his works was only the beginning of the early engine-builder's worries; he now had to machine them, and in the early 1800's with very few machine tools to speak of, this meant weeks of hard work with hand tools, everything depending upon the skill of the workman. The original lathe, the pole lathe, came down the centuries from remote antiquity and was the only type of lathe available until near the end of the eighteenth century. It was essentially a dead-centre lathe, the job being supported by fixed poppet-heads clamped to a single wooden beam. A cord at the end of a spring pole made several turns round the job and its end was fixed to a treadle, rotation of the job being towards the operator on the down stroke. This lathe was quite popular with woodworkers throughout the nineteenth century and is still in use in some parts of the world for such things as chair legs. With the passage of time, the pole and treadle were replaced by a hand wheel driving a rope round the job, thus ensuring continuity of rotation, and eventually the drive was made more positive still by water power.

Metal-turning was very much an art with these old lathes, before the slide rest came into being. Every cut had to be made by hand tool, as for wood-turning, and it was impossible to attain accuracy except by the expenditure of infinite patience: only an expert could hope to turn a truly cylindrical bearing. The cutting tool was fitted into a long bar which the operator held very firmly under his armpit, as for wood-turning, and in heavy work he was provided with a leaning post behind him against which he could steady himself. Each application of the tool on the job gave a circular cut, and a series of cuts roughed the forging down to near the required size. These cuts were then blended together into a relatively smooth surface, being finished off with a file and sanding block. In a heavy forging rotated by water power, hand-turning was not only difficult work, it was also very dangerous: any roughness in the forging skin could cause the tool to dig in and dislocate the operator's shoulder, or spring the job out of its centres on top of him. The surface finish on a crankshaft journal turned in such a lathe would be very coarse, suitable, possibly, for a slow-running paddle bearing lubricated with tallow and sea water, but hardly acceptable in a modern engine works. The as-turned finish would be readily measurable by hand calipers and steel rule and the tolerance would be in the region of ·01 of an inch, i.e. a surface that would be rough to the touch. By contrast, the surface finish in a modern crankshaft journal is specified in micro inches (one micro inch=one millionth of an inch) and the acceptable tolerance is from 10 to 16 micro inches. In some special cases, in precision pump spindles, for instance, the surface finish can be quoted as ·000003 of an inch, a mirror finish, in fact, and one only achieved by high-speed grinding, honing and lapping, but the basic shape is obtained by the use of machine tools.

Obviously there were no high-speed grinders in our forefathers' day, but neither were there any decent cutting tools. In the days before commercial steel came on to the market, and when metallurgy was in its infancy, the only cutting steel available was laboriously produced by the cementation process, a form of case-hardening by soaking soft puddled iron in a charcoal fire so that it could absorb carbon to enable it to be hardened and tempered. This sort of steel could not work at high speeds or with coarse feeds, the edge simply wore away. It had to be used on the slow-turning lathe of the period, barely pecking at the iron, because there was nothing else to take its place until carbon steel took over.

Nowadays, in our modern engine works production rate has become so important that the cutting speed and feed required would just burn the edges off ordinary lathe tools, and these are now made of high-speed steel. This material produced experimentally in 1860 by Mushet as a tungsten-manganese self hardening steel, was developed at the turn of the Century into an extensive group of alloys that would permit the tool tip to be run at a dull red heat.

There are five hardening elements used in the production of high-speed steels, the three more common being Molybdenum, Chrome and Vanadium, with two more exotic and expensive elements being used for the higher grades: Tungsten (W) and Cobalt. Thus, three or four of these, in an infinite variety of combinations, will produce a series of cutting steels with an increasing work/endurance capacity in relation to the percentage of hardening elements in their composition. These alloys are generally referred to by their formulae as a rough and ready guide to the purchaser, thus:

%W	%Cr	%V	%Co
14	4	0–1	—
18	4	1	—
22	4	1	—
18	4	1	5
18	4	2	10
22	4	1½	12

The great variety of these combinations is not necessarily a purchaser's whim. In these mass production days, where forming tools are often used and have to be sharpened in a jig, any time spent in dressing the edges means loss of profits, and tool changing must be carried out during, say, lunch break or at night. Thus a tool must retain its cutting edge for a precise period, which can only be decided in the shops, and the appropriate formula will then be quoted. Too good a steel will produce an edge that may last all day and well into the following, but may then need to be changed when the machine is supposed to be running.

Molybdenum can be used instead of Tungsten in the ratio of about 1%Mo. to 1½%W, and is, in fact, preferred in the Americas, where this element is mainly to be found.

The engineering industry took a very great step forward when Henry Maudslay, a prolific inventor of machines and machine tools, brought out his slide rest fitted on a lathe with cast-iron sheers in the last years of the eighteenth century. This tool rest had hand traverse and facing motion and for the first time ever it was possible to guarantee a truly round and parallel finish. He followed this innovation with a lathe where the work was held in a chuck and the slide rest travelled on double Vee sheers instead of the old-type round bar. Successive developments over the years introduced the self-acting slide rest, by Joseph Whitworth, and at long last the screw-cutting lathe, again by Maudslay.

The planing machine was a logical development of the lathe slide rest and did away with the necessity for all that laborious chipping and filing previously required to bed two castings together. The earliest planing machine, in the 1830's, had its table drawn back and forth along cast-iron sheers by means of a chain wound round a hand capstan, the tool box being hand reversed at the end of each stroke to take a cut on the return stroke. Some idea of the enormous amount of donkey work involved in building these early engines will be appreciated when one considers first the chipping and filing of the vee-shaped sheers for the plane table, and again of the team of men whose sole function in life was to haul round on the capstan all day, day after day.

As with metal turning, so it was with drilling holes. At first, the bowstring drill, a smaller version of the pole lathe, where the string made several turns around a wooden stock into the end of which was wedged a chisel—pointed drill. In passing, this early type of drill, coming down to us from the stone age, is still in use in some Mediterranean countries. The bowstring drill, being unsuitable for iron, was soon replaced by the drilling machine which was simply a table that could be moved up or down and a crank brace, pressure being applied by a weight bearing on top of the brace. Rotation of the brace was by hand power until well into the 1840's when Nasmyth introduced power drive and automatic feed, but the old diamond-pointed drill still held the field, generations had to pass before the twist drill took over. An outstanding feature of any Victorian marine engine, or any other type of engine for that matter, was the ubiquitous cotter, binding together the two halves of a bearing, wedging a stay into a casting or locating and clamping a crankpin in position. An apprentice could make an iron wedge from a piece of scrap metal and the cotter was a very simple solution to many engineering problems: it was also the only solution to most of them. Although machine-made screws had been turned out by instrument-makers in the eighteenth century, it took Maudslay's screw-cutting lathe to produce nuts and bolts in any quantity, and the early engineers had to make their own bolts and nuts. Until well after the start of the nineteenth century—in fact until about the time of the Battle of Waterloo—it took a craftsman to produce a threaded bolt. A piece of bar was first fashioned to the required shape and the thread was then laid off by winding a string round it, marking off the position of the spiral, then filing the thread into the bar. When the required length was roughed in, the bolt was returned to the lathe and the thread was chased or cleaned up by hand comb. As each engineering shop had its own standard-thread diameter and pitches, and nuts were not interchangeable, it was most important when dismantling any engine to secure each nut to its individual bolt, a practice that is followed to this day in any ship's engine room.

Nuts were produced in the same way as bolts, being forged square with a reduced hole punched through them which was then opened out to tapping size. A long taper was turned in a prepared screw of the right proportions and this was screwed as far into the nut blank as it would go, leaving a light impression of a thread in the bore, and this gave a start to the chasing comb to complete the thread in the lathe.

The actual power developed by these early marine engines was always an elusive quantity and any figure quoted should be accepted (as for a used car) with strong reservations. James Watt's formula for IHP was not always easy to apply in the years before the pressure gauge was invented, and boiler pressure was pure guesswork. In any event, most boilers were worked at less than 5 lb pressure and the engines ran on a good healthy vacuum. Again, revolutions per minute were not so easily come by when possibly the only man on board who possessed a watch was the captain: more guesswork. In land engines the efficiency was worked out in terms of duty, i.e. the amount of water pumped against the weight of coal burnt: a simple, straightforward equation which any mine-owner could understand, but at sea things were different; there were so many variables apart from revolutions and steam pressures: wind, weather, tide, loading, etc. Some special method had to be thought up to describe the dimensions of a steam engine for the convenience of builders, ship-owners and port authorities to enable them to compare like with like, something, in fact, similar to the Admiralty Coefficient beloved of Victorian ship designers, to compare performances of roughly similar ships. Such a criterion was the Nominal Horse Power, worked out by Boulton and Watt as follows:

$$\text{NHP} = \frac{7 \times 128 \times \sqrt[3]{\text{stroke in feet}} \times \text{piston area square inches}}{33{,}000}$$

or more simply:

$$\text{NHP} = \frac{\sqrt[3]{\text{stroke}} \times \text{dia.}^2}{60}$$

This assumed a general boiler pressure of no more than 7 lb.

The Admiralty formula was more realistic in that it took account of engine revolutions as well:

$$\text{NHP} = \frac{\text{piston speed in feet per minute} \times \text{piston area sq. inches} \times 7}{33{,}000}$$

or simplified:

$$\text{NHP} = \frac{\text{piston speed/min.} \times \text{dia.}^2}{6000}$$

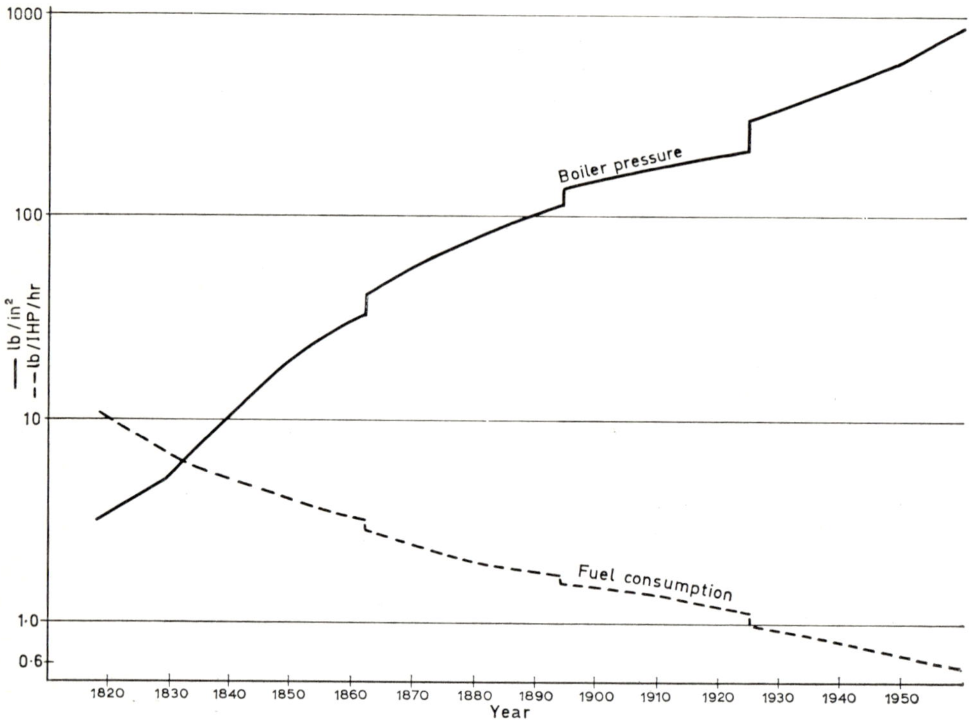

8 Fuel consumption related to boiler pressure (1820–1950)

Nowadays NHP is very seldom used, but when it is, it is taken as one-fifth IHP. With the early marine engines there was no clear-cut equivalent, but generally one will not be far out if one assumes the following ratios:

(a) Marine engines until 1820—NHP = IHP.
(b) from 1820 until 1840—NHP × 2 = IHP.
(c) from 1840 until 1860—NHP × 3·5 or 4 = IHP.

Closely related to the horsepower of an engine is its efficiency or, more usually expressed, its fuel consumption. Early land engines had no great problem here, as many of them were used in coal mines and could burn all the unsaleable coal they wanted right there at the pit head. Even in pumping stations outside the mining districts, coal of a sort was available, eked out by odd timber and rubbish. Although nobody wanted to burn coal unnecessarily, cost of fuel was the only worry: transport, access and bunkering were no problem at all. At sea things were different. The massive engine and the large, low-pressure boilers already took up far more of their share of space on board than was strictly warranted by their contribution to the ship's progress (until the 1870's steamers were still considered primarily as sailing ships), and the ship-owner objected to devoting most of the remaining space just to carry coal. Fuel economy was not an owner's whim, it was essential to the future of the steamship.

The first marine engines worked at a boiler pressure of from 2 lb to 3 lb, there being, of course, no expansion in the cylinder, and fuel consumption was, in consequence, extremely high, a reasonable figure being 10 lb coal/IHP/hour. With improvements in design, machining, alignment, heat insulation and boiler-making, allowing an increase in steam pressure, this figure was gradually brought under control, and by the 1830's we have a fuel consumption of 6 lb. This decreased slowly with the increase in steam pressures until, around the 1860's, compound expansion introduced a sudden step up in pressure with a corresponding step down in consumption. There is no such thing as an average steam pressure, but in the accompanying graph, relating boiler steam pressures to fuel consumption throughout the nineteenth century until the 1960's, the steam pressures and their corresponding fuel consumptions defining the curves were chosen from good passenger steamer figures. A point to remember: relating steam pressures of many hundreds of pounds with consumptions of less than unity in an ordinary graph would need an area the size of a blackboard to make the point. To get round this objection the figures have been telescoped on to logarithmic paper.

4 Paddle engines

We have seen how the pioneer steamers used paddles in their literal sense: a series of oars supported vertically on a framework and worked by a steam-operated mechanism to give them the to-and-fro movement. This mechanism proved too cumbersome and took up far too much room in the vessel, and subsequent inventors introduced rotary paddles, a series of 'oars' radiating from a rotating hub, and most illustrations of early steamboats show this type of undeveloped paddle-wheel, with four, six or eight spokes, each equipped with a rectangular float. These spokes were eventually wrought into a complete wheel with double rims to mount a certain number of fixed floats, the whole being constructed of wood assembled on to a cast-iron hub on the paddle shaft.

There was, and still is, no ideal position for the paddle-wheels in a ship. Some builders preferred side wheels, with the engines and boilers placed in a hold amidships, and this became the conventional position. In shallow-draught vessels, however, a single stern-wheel was preferred as being more efficient, with the engines and boilers on deck, while occasionally a ship was built with quarter-wheels, a pair of wheels one on each side of the stern, with the machinery in the after hold.

Contrary to popular opinion, the screw propeller is not necessarily more efficient than well-designed paddle-wheels, although it is certainly a neater drive to handle from the designer's point of view. A light-weight, fast-running diesel engine can be placed almost anywhere on board and coupled to a propeller either directly or through gearing, or, for that matter, by electric drive, whereas the paddle-wheels require large, heavy, slow-running machinery situated in the best and most useful part of the ship. But for shallow-draught vessels, paddles are still very much in evidence.

Whether paddle-wheels were made of wood or iron, it was the practice to fit wooden floats all round except for two or three opposite the crank, which were made of cast iron to balance up the running gear and prevent, as far as possible, the engine sticking on a dead centre, most engines in those days being single-cylinder units. For propulsion, the wheels should be situated near the crest of the bow wave to ensure constant dip or immersion, although it is doubtful if early shipbuilders worried over much on this academic point.

In a small steamer with low freeboard, where the paddle shaft is at deck-level, the floats enter the water at a relatively large angle to the vertical, and much of the engine power is wasted in forcing the water downwards at entry and scooping it upwards upon emerging. One solution would be to increase the wheel diameter, thus ensuring a better float entry angle, but to maintain good immersion, this would require the paddle shaft to be raised considerably above deck-level with possibly severe side effects on stability and convenience. In really large paddle ships, with wheels of 40 feet diameter and over, the paddles were always radial there being no point in making them otherwise, but in the majority of paddle ships, of small to medium size, with small fast rotating wheels, a neater solution was to feather the floats—to cause them to enter and leave the water in a less oblique manner than if they were fixed to the spokes. Feathering floats are hinged upon the wheel rim and are provided with an arm or crank connected by link gear to some part of the adjacent ship's structure whereby, on rotation of the paddle-wheel, the floats are moved relative to the rim.

Before feathering floats really took hold, however, several designs of fixed floats were brought out in an attempt to find a better type of float than the existing flat, rectangular one. Some floats were diamond-shaped, commonly called duck-foot boards, where the pointed end was to reduce the impact upon the water. Other floats were placed obliquely across the wheel rims, the obliquity of the port floats being opposed by that of the starboard ones. Possibly the best known of all patented fixed floats was Galloway's cycloidal wheel, originally introduced, but not patented, by Joshua Fields in 1833. In this wheel each float was cut into four horizontal sections, each section being mounted one behind the other like the risers in a flight of stairs, from the periphery towards the centre.

Feathering floats, first introduced by Buchanan of Glasgow in 1813, became established only about the 1830's, but by the 1840's there were many dozens of patents for feathering wheels; as is so often the case, the final design known as Morgan's wheel, which can still be seen in some of our existing paddle steamers, was based on Buchanan's original ideas.

The design consisted of a wrought-iron wheel mounted on a heavy cast-iron hub, with a number of iron floats equally spaced around the rim, independent of the spokes. These floats were hinged upon the rim and the relative movement was obtained by radial rods connecting the float arms with a feathering strap riding on a sheave fixed to the outboard side of the paddle box, level with the centre line of the paddle shaft and from 9 to 12 inches forward of it. This eccentric strap was driven by a driver, a radial rod made stronger than the rest and firmly bolted into the strap. Sometimes a fixed pin was fitted into the sponson instead of the sheave, with the same off-set centres.

Another type of feathering float derived its movement from a stationary crank inside the hub, where the crankthrow was slightly less than the float arm, thus giving a very efficient feathering action but a very weak paddle drive owing to the complication of the hub.

A neater type, quite common in its day, was Pool's patent as developed by Seaward and other famous engineers of the early nineteenth century. Here the floats were hinged about the wheel rims as usual, but their arms were equipped with small leather rollers running in a circular metal track built against the inside of the paddle box, this track being eccentric to the paddle shaft centre by the required amount to give good obliquity.

A more unusual type was one where each float revolved about a spindle athwart the rim, the rotation being in the ratio 2:1, the floats revolving twice about their axles for each revolution of the main wheel. By careful adjustment of the gears, the floats were in a suitably vertical position at entry, near the bottom of the wheel, while being almost horizontal at the top.

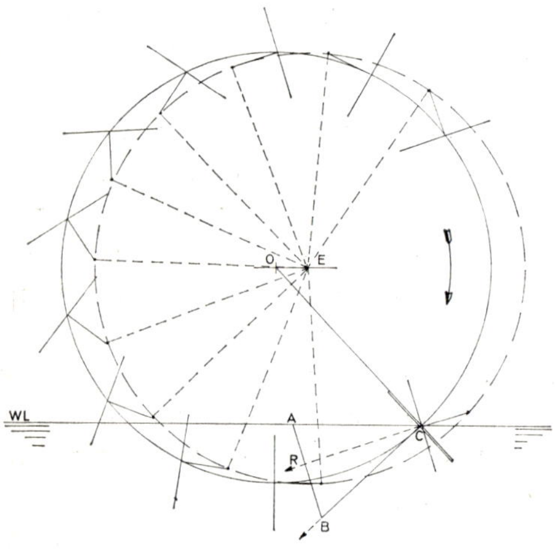

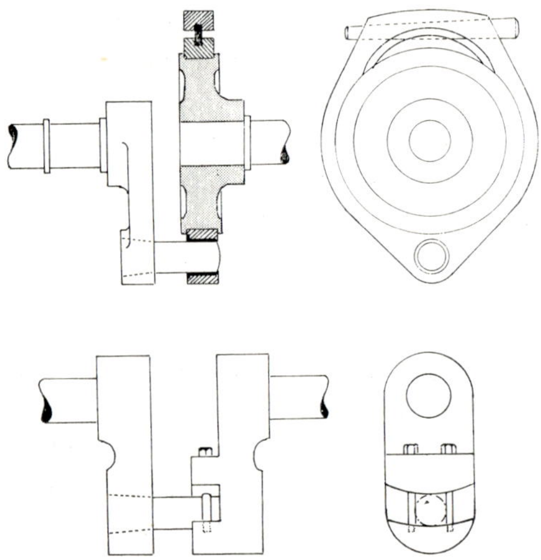

9 *Left:* Diagram of paddle-wheel showing mechanics of the feathering float
10 *Right:* Two types of gear for disengaging paddle-wheels

In the accompanying sketch of a feathering wheel, O is the shaft centre and E is the eccentric centre, and the effect of this eccentricity on the floats is to make the angle of entry roughly equal to a radial float on a wheel of twice the diameter. (Note: angle of entry is given by the float when immersed to its centre of pressure C.) When designing a wheel, to determine whether to use radial or feathering floats, it was sufficient to lay off on the water line CA = speed of ship in feet per min., and CB, drawn perpendicular to radius OC, equal to speed of paddle rim to same units; joining AB, we get the ideal angle of float obliquity which can then be transferred to point C to give ideal entering conditions. In a sufficiently large wheel this angle would approximate to that of a fixed float, while in a somewhat smaller wheel it might not be too far out to bother with the complications of feathering floats. The line CR, normal to the float at its centre of pressure, shows the direction of thrust and should be as near to the horizontal as possible.

Before dealing with a detailed examination of the various types of paddle engines, it will be necessary to look at some features common to all of them. The main driving shaft, for instance, consisted of three sections: a paddle shaft on each side carrying the wheel, and the intermediate shaft driven by the engine. In the relatively few ships where the paddles had outer bearings on the sponsons, ie where the wheels were supported on both sides, the paddle shaft was made parallel or, sometimes, tapering slightly towards the wheel. Generally, however, with an overhung wheel, the paddle shaft was made heavier at the outer end to take this considerable bending moment.

The intermediate shaft (what we would nowadays call the crankshaft) consisted of a plain shaft with a forged crank in the middle for the air-pump drive, and a shrunk-on overhung crank web for each cylinder. This web had the crankpin bolted into it, or driven in on a taper and locked with a cotter. The paddle shaft carried the mating web with a brass bush in the eye to receive the end of the crankpin: there was no continuity of forging as with a modern crankshaft. Occasionally, instead of the crankpin running in a brass bush in the opposite web, the end was made spherical, to allow the webs to adjust themselves to any misalignment of the shafting as a whole. Due to the weight of the overhung wheels, running in sea-water-lubricated bearings inside the paddle box, it was impossible to prevent heavy weardown in the shaft bearings. Also, as all of the thrust in a paddle shaft is taken up on the forward side of the plummer blocks, these bearings wore oval sideways as well as vertically. To prevent serious damage, the crankpins were made quite an easy fit in their outboard webs to allow for normal wear.

Another point which does not seem to be sufficiently appreciated is that throughout almost the whole of the nineteenth century ships of all types were designed as sailing vessels first, and their engines were there to be used only when the wind failed. Even the *Great Eastern*, that giant of the century with both paddle and screw engines, was still rigged for sail and she used her canvas for several years until her owners acquired confidence in the engines. Now it was claimed as proven fact that a paddle steamer in a storm had to keep her engines going, for, although the waves would smash the paddle boxes to matchwood, the moving wheels were somehow immune to damage from the seas. Once they were stopped, however, the sea took over and stripped them of their blades.

But in a fair wind the engines would be stopped and the vessel could proceed at a reasonable speed on sail alone. This was rendered the more necessary as bunkers were not available at every port and coal had to be eked out for the whole passage. To avoid drag from the paddle-wheels when under sail, some means had to be found to disconnect them from the engine to let them revolve freely or trail, and several novel ideas were brought out and tried at sea. The earliest was the reefing wheel, whereby about half the number of spokes could be gathered together like the ribs of a fan and lashed clear of the water, but this was rather a dangerous operation except in a dead calm. Another system was that of unshipping the paddles completely and stowing them on deck, with similar disadvantages.

Small coasters sometimes used dog clutches on the paddle shafts to engage the wheels, but this was impossible with a large heavy wheel. On some larger vessels the paddle shaft was made with a large axial clearance, such that it could be jacked outboard for about 5 inches, enabling the crankpin end to withdraw from its opposite crank to allow the wheels to turn freely. Two common methods in use are shown in the sketch. In one arrangement the paddle-shaft crank web is replaced by a cast-iron disc with a brake band encircling it. The crankpin engages in this band, which rotates with the engine. By knocking in the cotter, the band is jammed hard against the disc and drives the wheel. The claim was made that the brake could be engaged or disengaged by a few strokes of a maul, and the wheel could be released in a matter of seconds. The second method, whereby the crankpin is held in a circular groove in the web by a bolt on each side, would appear to be rather noisy in action, and releasing the pins could be dangerous for the operator, who never knew when a wave might strike the wheel and set it in motion, but, then, there is nothing inherently safe even in a modern engine room.

Another feature in common with most paddle engines and all earlier ones was the single eccentric pulley, set loose upon the shaft and fitted with a heavy counter-weight to balance the valve gear. This sheave was machined with an abutment that rode against one of two stops set into the intermediate shaft, one for ahead, the other for astern running. The eccentric rod drove the valve spindle usually through bellcrank motion, and it was made with a notch or gab which engaged in a pin on the valve gear. To stop the engine, the eccentric rod was simply lifted off this pin and the slide valve remained where it was, thus braking the engine on compression. To restart the engine in the opposite direction the slide valve had to be moved by hand, a fairly simple matter in a small engine where a hand lever was sufficient, but a very heavy, very dangerous job in a big engine with heavy valve gear.

The starting gear in the larger engines consisted of a large iron wheel similar to a ship's steering wheel with outside spokes, on one end of a shaft, with a pinion at the other end. With the eccentric rod released, this shaft was pushed forward and the pinion engaged in a rack on the valve motion; with a team of men climbing round the starting wheel, the valve would be gradually moved to the astern running position against the vacuum. As soon as the engine commenced to move, it was imperative to withdraw the starting wheel as the eccentric would immediately take up its astern position and drive the eccentric rod with all the power of the engine behind it; and once the gab dropped on to its pin on the valve motion, the valve travel would be automatic. If the starting pinion were still engaged in its rack the heavily spoked starting wheel would become a murderous instrument.

In twin-cylinder engines separate starting wheels were fitted to each slide valve, and in manœuvring it was necessary to keep a watchful eye on the starting team on the other side of the engine room, as well as on one's own side. Not only was it necessary to have a team of men able to withdraw the gear quickly, as one man, as soon as the engine started, there was very little time margin to make up one's mind. If the valve did not move in phase with the shaft revolving in the new direction the engine would choke and stop again, and the valve would have to be laboriously moved over again by hand for a fresh stroke.

In later paddle engines this danger was appreciated and a special device was incorporated into the reversing/starting gear. By pulling up a lever mounted on a bellcrank, the eccentric rod gab was disengaged from the valve pin, whilst, with the same movement, the hand gear was coupled to the valve spindle. By letting the lever drop the hand gear slipped free and the eccentric drive re-engaged with the valve spindle as the eccentric moved over. It will be noted that in these low-pressure engines the steam was not shut off for manœuvring, it was considered sufficient to disengage the slide valves.

Although the tubed condenser was patented quite early in the nineteenth century, in 1839, very little use was made of it; marine engineers favoured the well-tried jet condenser, whereby the exhaust steam from the cylinders was condensed directly by contact with sea water, most of the condensate being discharged overboard with the exception of a small amount pumped back into the boilers as feed. The jet condenser gave a very good vacuum, the sea water being introduced into the condenser from a Kingston non-return valve on the ship's side through a nozzle, the jet usually impinging on the wall of the air pump, which sat inside the condenser, to form a sheet of spray.

This jet condenser was also used to pump out bilges, and there was no serious objection to this as long as the air-pump valves were not choked up. The air pump, in fact, was used in the earlier engines as a donkey pump for pumping out

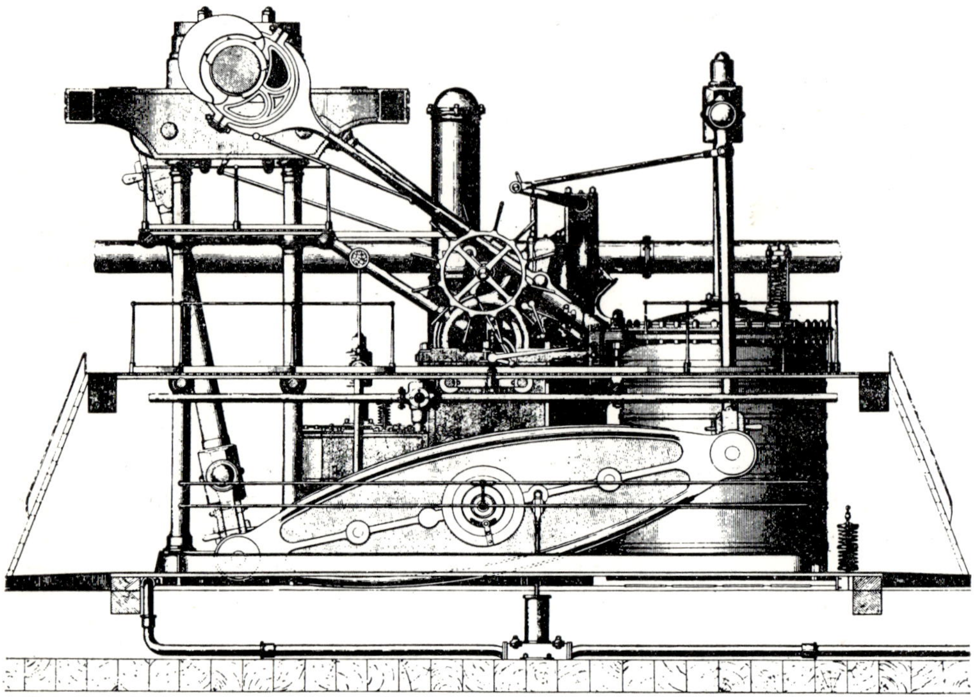

11 Side-lever paddle engine (c. 1860)

condenser, bilges, pumping up boilers or water on deck: it was simply *the* pump. No tubed condenser could hope to stand up to this sort of treatment, especially in view of the prevailing fashion of lubricating the cylinders and slides with an abundance of tallow injected into the steam spaces through a tallow kettle. In passing, it might be mentioned that the modern emergency bilge valve on all ships, commonly called the bilge injection, originated in the practice of condensing the steam by drawing in a spray of water from the bilges instead of direct from the sea.

The one serious drawback to the jet condenser in the days when sea water was used in the boilers was the danger of flooding the engine. To start, or reverse, one of these huge lumbering paddle engines with steam pressures as low as 5 lb or at most 20 lb required a good vacuum, and to obtain this the engines were filled with steam until it issued from the snifter valves on the condenser. The sea-water spray was then smartly turned on by the regulating cock on the platform and a vacuum thus formed. In manœuvring from ahead to astern it was essential to station a man at this cock to shut off the sea water and reopen as required, or the engine flooded hopelessly and the condenser had to be drained by hand. Alternatively, if the flooding passed unnoticed in the scramble to get the engine turning, something unpleasant happened to either the air pump or the steam cylinder.

In modern times, when an ordinary marine engine can be easily handled by one man, a description of manœuvring a large paddle engine in the 1850's makes interesting reading: at standby, the watch below was called out and, together with the duty watch, took their stations at the various controls—three men to each starting wheel, one man to each of the port and starboard condenser discharge valves and one man to each injection cock, with the engineer in charge directing operations. Meanwhile, the remaining members of the engine-room crew attended to routine greasing, water-tending and stoking.

Now, as regards the early paddle engines, there were as many kinds as there were marine engineers to build them; there was no standard design as in the modern marine engine, and every engine-builder designed to his own whim without too much regard to proportion, weight or strength. Even so, the various engines came under two different categories: indirect and direct drive. The indirect drive depended upon a beam to transmit the power to the crank: first the overhead beam as with land engines, later converted into an underslung beam or side lever which also included the grasshopper type and the bellcrank levers.

The direct-drive engines were divided into five different groups or classes according to the way the stroke of the piston was conveyed to the crank. The Gorgon engine, which eventually became the conventional present-day marine steam reciprocating engine, had the connecting rod between the piston-rod crosshead and the crank bearing. The steeple engine, or return-connecting rod engine, had the crankshaft placed between the cylinder and the crosshead. The double-crosshead design stemmed originally from the side-lever type, with side rods from the piston yoke driving a pair of side crossheads which in turn drove a forked connecting rod on to the crank. The Siamese engine consisted of a pair of identical cylinders with their pistons acting together on to a common connecting rod. The principle behind the steeple, double crosshead and the Siamese engines was to obtain as long a connecting rod as possible with the cylinder tucked below the intermediate shaft. The fifth engine, the oscillating engine, evaded the problem of the long connecting rod by dispensing with it altogether. Here the piston rod was coupled direct to the crank, the cylinder rocking back and forth on trunnions to accommodate the crank rotation.

Without a doubt the most popular paddle engine of any class and at any period

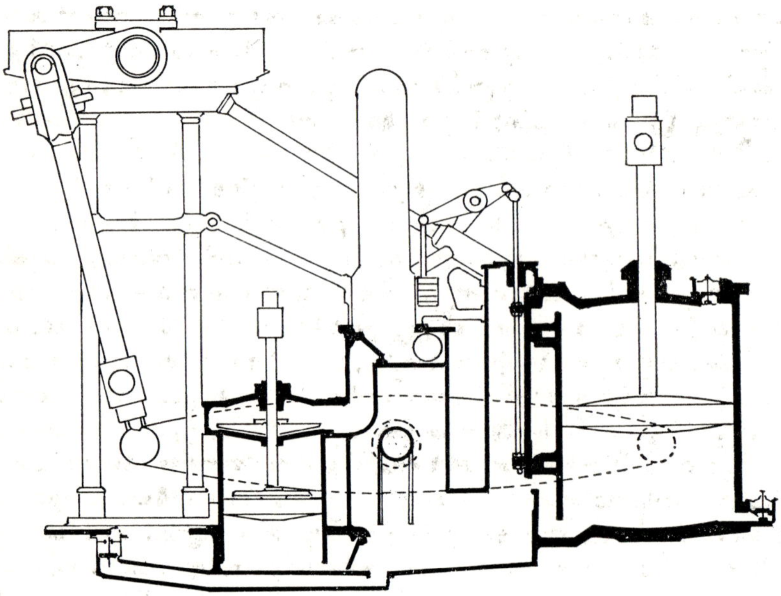

12 Section through a side-lever paddle engine

was the side-lever type, sometimes with two cylinders, but more generally with a single cylinder, or, in the phraseology of the day, built as a single engine. By the 1850's this type had crystallised into a formal design in general use throughout the whole spectrum of merchant ships, from the homely tug to the Atlantic passenger liner. Mechanically and aesthetically, it was a beautifully balanced engine, set upon a square box bedplate.

The jet condenser occupied the centre of the bedplate and it contained the seating for the vertical steam cylinder at one end with the air-pump cylinder embedded into the other end, the three units: cylinder, condenser and air pump being in direct communication with each other by slide valve and foot valve. The main side levers were mounted on a rocking shaft taken athwartships through the condenser in a conduit, and which was supported on pillow blocks bolted to the bedplate. Motion from the piston-rod crosshead was transmitted to one end of the levers by side rods, and from the other end to a crosstail by short links. The connecting rod had its lower end driven into the centre of the crosstail on a taper, and its upper end was coupled to the crankpin.

The framework consisted of a cast-iron entablature containing the main shaft bearings supported by either heavy cast-iron Gothic columns or more

13 Parallel motion mechanism for a rotative beam pumping engine (1870)

generally by a light-weight structure in wrought iron, with diagonal stays to the top of the cylinder.

The disposition of the engine with a vertical cylinder allowed no room for the crosshead guide, and side thrust was taken up by the parallel motion, a pair of horizontal links swinging from the crosshead coupled to a pair of vertical rods worked from some point on the beam centre line such that, with the side rods and the portion of the beam, they formed a parallelogram. To prevent this parallelogram from collapsing, a horizontal radius rod of predetermined length and anchored to the cylinder cover was connected to a pin on the vertical links. As long as the linkage in the parallel motion was kept well adjusted, this type of guiding gave no trouble at all.

The motion just described was that introduced into the marine engine by Boulton and Watt; there were many forms of parallel motion all designed to guide their focal point—the piston-rod crosshead—in a straight line, and most of them were accurate enough for the slow-moving engines of the day even if they were not mathematically correct.

The air pump had no system of group valves as in the present-day pumps. A single hinged-flap valve at the suction from the condenser, a pot-lid valve in the bucket and another of the same type in the discharge cover. There was usually another non-return hinged valve in the hot well above the air pump, leading to the waste-water pipe overboard and the feed-pump suction.

The side levers, originally made of wrought-iron plates, eventually came to be cast in iron. The levers and their articulations gave no trouble, being simple and strong, but the weak point about all side-lever engines was the fulcrum shaft and its bearings. These took the combined reactions of piston and connecting rod, plus the considerable weight of the beams themselves, of the side rods and the linkages, air-pump drive and the massive crosshead and crosstail forgings. When the ship rolled, the levers bore hard on their bearing collars which tended to overheat, in spite of the tallow spread over them. Also, as the fulcrum shaft was carried through the condenser it involved either a tubular core in the casting to accommodate it, or building the shaft into the condenser, with a stuffing box on each side with the probability of air leakage later on.

The valve gear as already described was of the single eccentric type with balanced sheave and disconnecting eccentric rod, with the valve worked by lever or by handwheel when reversing and starting. In the paddle engines under survey, which were built up to the 1860's, some of them attained 4000 IHP on two cylinders, at steam pressures of from 20 to 25 lb. Thus a steam cylinder developing some 2000 IHP on steam at 25 lb was a monster of a cylinder indeed, with a slide valve to match, and, in manœuvring, this valve had to be worked by hand and controlled to a split-second timing. Any inattention on the part of the men operating the starting wheel could mean serious bodily injury either from the engine as it started up on automatic drive or from the engineer in charge.

Finally, a very small item of major importance on any jet condensing engine: the snifting valve. This was a non-return valve from the bottom of the condenser which served to drain an accumulation of water in an emergency and so protect the pump, but also used to expel the air when blowing the engine through with steam to establish the vacuum; without a suitable vacuum it is doubtful if many of these old engines could have worked at all.

A variant of the side-lever engine was the half-lever or grasshopper engine. This was, in fact, the earliest form of indirect drive to be used on this side of the Atlantic and it may be recalled that the *Comet* (1812) was fitted with a grasshopper engine. It was also the latest type to be used, lingering on until the 1920's and possibly the 1930's. As this type was being built long after screw engines had become established and their design refined, so the grasshopper engine in later

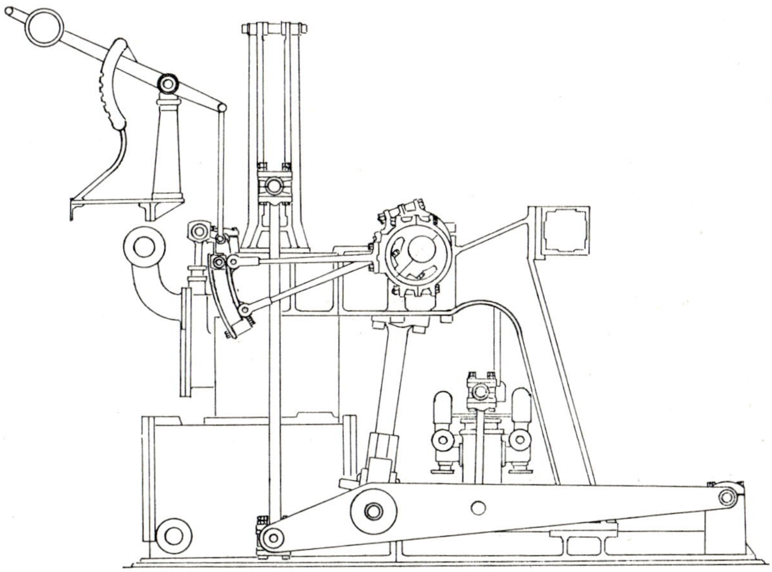

14 Half-lever or Grasshopper engine

years showed features not usually seen on the older paddle engines: crosshead guides, either shoes or rollers, in a special track, and double eccentrics with link motion driving the slide valves.

A third type of side-lever engine, popular enough in the earlier years of the marine steam engine but discarded round about the 1830's, was the bellcrank drive, either singly or with two cylinders placed side by side on to the same crank. It was an uncouth type of transmission, neither pretty nor practical, and it is hard to see how it was ever introduced to replace the established overhead or underslung beam.

Now in all cases of paddle drive the engine-builders' main preoccupation was the length of the connecting rod and how to fit it into the space available. The connecting-rod problem acquired a mystique all its own and its length has been handed down to us as not less than four crankthrows. In the side-lever engines and their associated types the connecting rod could be made almost the full depth of the hold, from the crosstail near the keelson to the intermediate shaft on deck —there were no problems here. But in the direct-drive engines things were not quite so easy. The pioneer steamboats, being small and with no depth of hold worth speaking of, had their machinery and bunkers sprawled all over the length of the ship, with no room whatever for passengers or cargo. Even in later years, as

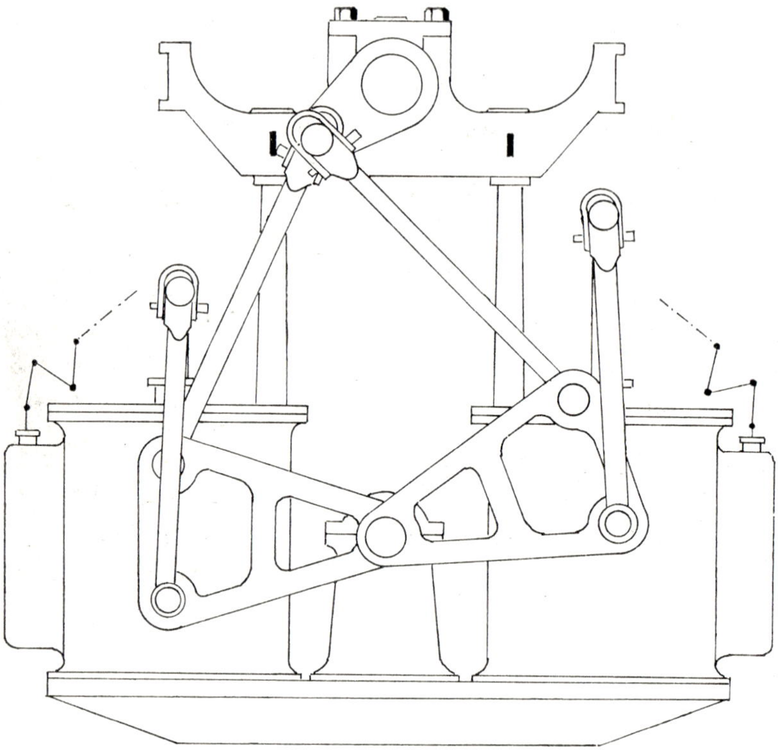

15 Bellcrank lever paddle engine

the engines became more and more sophisticated, the machinery occupied far too much space in relation to its value in a profit-making venture, for it must be remembered that until the advent of compound expansion, and the reduction of fuel consumption to a reasonably economic figure, steam had great trouble in holding its own against sail, where the wind was free.

In direct-drive engines, therefore, only a limited engine-room space was offered to the engineer, and even in a smallish steamer a horizontal engine took a lot of space. Again, the Merchant Navy, which had taken and kept the initiative of steam propulsion and engine design ahead of the Royal Navy, took its cue from the latter with regard to the cylinders, which had at all costs to be kept below water-level out of reach of shot. With these two requirements in mind, cylinder near the keel and as near as possible to the shaft, the only position left was to site the engine vertically immediately below the shaft.

One of the earliest types of direct-acting engines was Seaward's Gorgon

16 *Left:* HMS *Gorgon* (1837)—Early version of direct-drive paddle engine
17 *Right:* SS *Helen MacGregor* (1843)—Inverted Siamese engine

engine, so-called from HMS *Gorgon*, built in 1837. All subsequent engines of this type were invariably referred to as Gorgon engines until Thomson turned it upside down, called it the steam-hammer engine and used it to drive a screw.

The *Gorgon*'s twin-cylindered engine was remarkable for being exceedingly squat. The cylinders were placed vertically square on to the bedplate and the cross-heads were coupled to the cranks by very short connecting rods, the ratio of connecting rod to crank being little more than 3 to 1. Guide shoes and plates were not provided; instead the side thrust on the connecting rod was taken up by heavy parallel motion, part of which incorporated a set of rocking beams to drive the air and feed pumps. The jet condenser was placed in the centre of the bedplate between the steam cylinders and the two air pumps.

The valve gear for each cylinder was worked by a single eccentric balanced sheave and disconnecting rod. A rather unusual feature, however, were the two small distribution valves, one for steam, the other for exhaust, in place of the usual large slide valve. This gave separate control for the steam and exhaust valves at each end of the cylinder: the eccentric rod gave movement to the steam-valve spindle and this in turn, through an adjustable bellcrank, worked the exhaust-valve spindle.

The early Gorgon engines were not well received—engineers disliked them, claiming their short connecting rods caused overheating and wear in crosshead and crankpin brasses. Also that they caused too much wear on the parallel linkage pins. In short, the engineers of the day, being used to the more spacious arrangements in the side-lever engines, disliked having to stow the whole of this direct-action engine in the small space between keelson and deck. It was not until diagonal engines were used to drive the paddles that the Gorgon engine really came into its own.

The engines of the *Gorgon* were certainly not the first of their class—the *Charlotte Dundas* had direct-drive engines, placed horizontally, as also did the *Savannah,* but the most important of these early 'conventional' engine types were those of the Arctic explorer Captain John Ross's *Victory*. Although the engines drove direct through a connecting rod on to the crank, the paddle shaft was driven on the 'second motion', i.e. the paddle shaft was geared down from the crankshaft. In this particular ship each paddle could be worked independently by its own engine.

The machinery was designed and supplied by Braithwaite and Ericsson and the very odd feature about this installation was the use of a main and auxiliary blowing engine to provide a draught for the boiler furnaces, thus dispensing with a funnel. The bellows was simply a wood-and-leather blowing cylinder with three pistons in tandem working in a series of short chambers. As was seen in the short description of the *Victory*'s voyage to the Arctic, this engine was not a success and must be chalked against this gifted inventor's wonderful record.

Having registered their protest at the Gorgon engine's short connecting rod, the early Victorian engineers immediately set about designing some alternative form of drive wherein the length of the connecting rod would not be tied down to the small space between cylinder cover and crankweb.

Maudslay solved the problem with his double-cylinder or Siamese engine, there being two identical cylinders to each crank. The two pistons rose and fell simultaneously as one piston, each piston rod being connected to the opposite end of a T-shaped crosshead, the perpendicular part of which fitted between the two cylinders with a guide shoe at the bottom engaging in a guide track on the cylinder walls. A connecting rod extended from the bottom of this T crosshead to the overhead crank, the length of this rod being related to the depth of hold.

Maudslay's Siamese engine may have solved the problem of the connecting-rod length, but it introduced additional problems: duplication of engine parts, greater condensation and heat leakage, extra weight, *et al.*

However much the Siamese engine may seem odd to modern engineers, Forrester's variant of it as fitted in the paddle steamer *Helen MacGregor*, built in 1843 for the Hull–Hamburg service, was very much odder, in fact a freak. This was an inverted Siamese engine with the bottoms of the two cylinders close against the intermediate-shaft entablature and the piston rods pointing vertically downwards through double stuffing-boxes with oil-retaining glands. The two piston rods fitted into a common crossbar extending below the two cylinder covers, with a heavy cast-iron vibrating frame forming the parallel motion and serving to work the air, feed, bilge and brine pumps by means of rocking levers. The connecting rod was connected to this crossbar and drove the single crank direct. Each cylinder had its own slide valve, the two spindles being worked by a bell-crank from a common single eccentric sheave, while the air pump and jet condenser were as for ordinary installations.

There seemed to be no special merit in this particular engine, apart from its compactness. The builders' claim that it was impossible for sea water to flood the steam cylinders because of their height above the keel line was negatived by the proneness of the air pump to choke whenever the ship rolled.

A different approach to the connecting-rod problem was made by both Bury and Fawcett in their double-crosshead engines. In this type, usually a single (cylinder) engine, the cylinder was vertical and the crosshead formed a crossbar extending the full width of the cylinder, from the extremities of which a couple of side rods extended, as in the case of a side-lever engine. The lower ends of these side rods were guided along the walls of the cylinder, one on each side, and from their extremities another pair of side rods ascended to a crosshead situated above the crossbar, from the centre of which the connecting rod proper drove the crank. This was, in fact, the equivalent of a forked connecting rod whose jaws spanned the cylinder diameter to reach the side rods.

Although quite a number of Siamese and double-crosshead engines were built and successfully run, they were never popular with the ship-owner of the day, and it required an engineer of Napier's calibre to get the direct-acting engine established. This he did with his steeple or return-connecting-rod engine, where the cylinder sat on the keelson with its cover close to the intermediate shaft and having the advantage of a long stroke. The piston had two or four piston rods passing on either side of the shaft and extending to a crosshead with rollers running in a guide race on a steeple-shaped structure high above the deck. From this vantage point the connecting rod drove the crank, its length being unhindered by

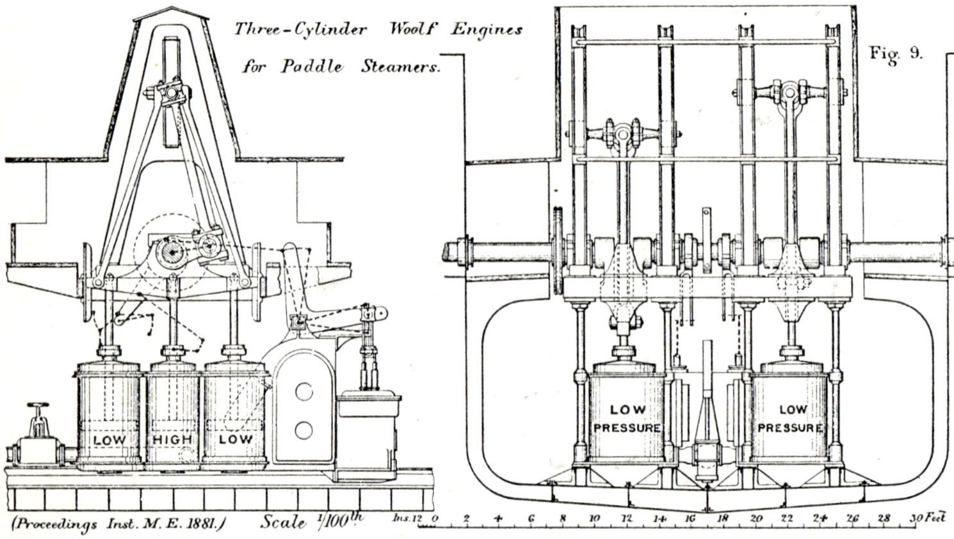

18 Napier's steeple engine

other practical considerations. The piston valve was worked by the usual single eccentric and the air pump by linkage from the crosshead and rocking beams.

Although David Napier's engine came to be accepted for river craft, where paddles alone were used for propulsion, it never became popular with deep-sea vessels, where steam was often an auxiliary to sail. Although grateful for the benefit of that extra foot or so of connecting rod, the ship-owner felt he was providing an awful lot of engine for the amount of power developed; an engine, moreover, which protruded through deck and crank hatch and stuck out at all angles to the discomfiture of his deck officers. The steeple engine proper disappeared with the passing of the paddle engine, although it lingered on for a while as a horizontal low-pressure return-connecting-rod engine for driving the screw. The principle, however, lives on in the banjo donkey pump which is still to be met with on old coastal steamers in tropical ports.

This intense preoccupation with the length of the connecting rod led early Victorian engineers to design such extravagant types of engine as the Siamese and the steeple drives without pausing for a moment to see if possibly there was an alternative solution to the problem: a logical approach rather than a purely mechanical one, by dispensing with this troublesome item altogether. The principle of the vibrating engine had been put forward by Murdock as long ago as 1785 and patented by Witty in 1811, and, it may be recalled, an oscillating engine propelled the *Aaron Manby*, built in 1822. Joseph Maudslay advanced the design and helped to market the engine, but it was many years before Penn succeeded in establishing the oscillating engine, probably as much by sheer excellence of workmanship as by any other quality in the engine itself.

The original oscillating engine had the trunnions placed fairly low in the cylinder, giving little sway, and the valve chest was fitted on the after side, at right angles to the trunnion axis. The valve spindle was worked by a forked arm on the crankpin bearing striking against upper and lower stops on the spindle, rather in the manner of a steam-pump valve. This valve movement, depending as it did on operation at each end of the main piston stroke, was suitable for a very low-pressure engine only, as no steam expansion was possible.

Maudslay's oscillating engine as supplied to the steamer *Endeavour* in 1828 had a fairly simple cylinder set on trunnions and swaying on a pair of triangular frames with the shaft bearing at their apex. This cylinder had the slide valve at the side, in line with the steam trunnion, the exhaust being carried around both sides of the cylinder in a wide belt to the exhaust trunnion, thence to the jet condenser. Valve motion was supplied by the usual single eccentric with the eccentric rod end guided in a slot in the cylinder wall. From this guide shoe a frame transmitted the movement to the valve spindle.

John Penn's engine was very much more sophisticated in design. The cylinder trunnions were fitted to the cylinder walls at such a height that, with the piston on bottom and the crankpin bearing uncoupled, the cylinder would not capsize. In some cases the trunnions were cast separately and bolted on to a machined recess in the cylinder walls. By now the engine had two slide valves per cylinder, one forward, one aft, with a wide belt encircling the cylinder waist, for steam to and exhaust from the chests. Steam from the main boiler stop valve (there was seldom a manœuvring or throttle valve, the engine could be stopped just by disengaging the eccentric rod from its pin) entered this belt through the outboard trunnion in a twin-cylindered engine, and exhausted straight to the condenser through the inboard one, the trunnions working in large bearings seated on pillow blocks with expansion joints and stuffing boxes at the steam-pipe connection.

The valve motion was rather complicated inasmuch as two valves had to be operated by a single or double eccentric linkage in such a manner that the swaying of the cylinder would not affect the timing. The illustration shows a two eccentric arrangement where the reversing links engage a fixed block on the main quadrant. This quadrant reciprocates in harmony with the eccentric rod, and is guided by shoes on the main entablature pillars and by a tail rod. The main quadrant is curved to an arc struck from the centre of the cylinder trunnion with the valve

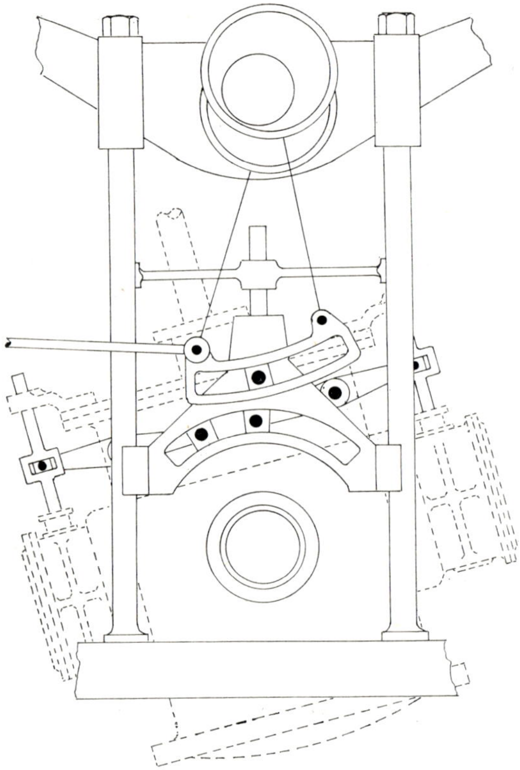

19 Valve gear for an oscillating engine

at half-stroke, and it carries two little sliding blocks attached to the ends of the valve rocking levers. By this means the movement of the eccentric can be transmitted direct to the valve spindles to give correct timing whatever the inclination of the cylinder.

In the case of a single eccentric reversing engine the quadrant remains as already described but has a vertical rack cast on it, meshing with a pinion and shaft, the other end of which carries the starting wheel. This pinion is engaged for starting, when the valve spindles can be worked by hand and withdrawn as soon as the engine picks up under steam.

In the oscillating engine the tendency was for the cylinder and piston-rod stuffing box to wear oval, and this was countered by fitting a fairly deep piston and sweeping the cylinder cover up into a cone shape with a very deep stuffing box. In general the combined depth of neck and gland bushes measured half the cylinder diameter or more. The trunnion bearings gave very little trouble: with boiler pressures of 25 lb or thereabouts and a steam temperature of little more than 220°F in the cylinder, the temperature of the inlet steam at these bearings was sufficient to keep them running at safe working heat while the exhaust side remained cool at all times, although here it was important to keep the trunnion gland tight.

If the oscillating engine had a weakness at all it was in the air-pump drive. Side-lever engines and direct-acting engines usually had this important pump drive either from some point on the side levers or by linkage from the piston-rod crosshead and a reasonable stroke could thus be obtained. In the vibrating engine the action was from a forged crank in the intermediate shaft: a very difficult and expensive shaft to produce and one which of necessity could provide a limited stroke only. For a proper understanding of the paddle engine it is necessary to keep the air-pump function very much in mind as this was the key feature in any early Victorian engine. The capacity of this pump had to be equated exactly with the injection cock opening; if the engine slowed down due to stress of wind or weather the air pump slowed down with it and the injection *had* to be reduced to avoid loss of vacuum, flooding, engine stalling or cylinder damage. With a short air-pump stroke it would be more difficult to catch up with flooding than in the case of one with a reasonable stroke.

Feed and bilge-pump rams were normally driven by a large eccentric bolted to the trunnion giving a short steady stroke, without too much shock to the feed piping and check valves.

There were many different makes and fashions in direct-acting engines during the heyday of the paddle steamer, 1810–70, but they all fell within the five categories already mentioned. Of the many thousands of all types built during that period, two types only need be considered: the side-lever engine, with its cousin the half-lever or grasshopper engine, and the oscillating engine. These two were the only ones to outlive their century and to survive well into the 1900's. Both were in evidence in the 1930's and possibly even later.

Towards the end of the century, paddle engines were built as diagonal conventional engines with compound or triple expansion, but in all cases the type was kept pure for all cylinders, with the following odd exception which remains an interesting hybridisation of types. This was the Blackpool–Isle of Man steamer *Queen of the North*, built in 1895 at Birkenhead, where the intermediate shaft was driven by four cylinders on two cranks, the HP cylinders being of the conventional Gorgon diagonal type one to each crank and placed forward of the shaft, while the LP cylinders were oscillating, placed immediately below the shaft, with the HP and LP crankpin bearings working side by side on the pin.

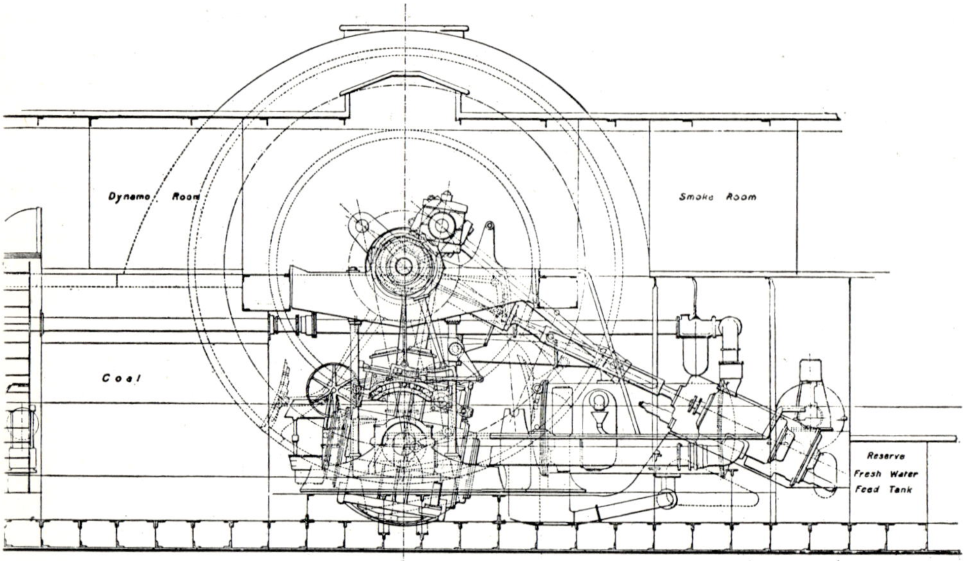

20 SS *Queen of the North* (1895)—Combined oscillating and direct-acting paddle engine

An even more unusual paddle installation was that once seen on the screw steamer *Ardenrigh* some time in the 1880's. This vessel sustained severe damage to her propeller and shed all the blades, leaving only a useless boss. The captain rigged such canvas as he had on board but the weather remained a dead calm and, being outside the steamer route, he had no chance of a tow. In desperation he decided on a jury paddle rig to find his own way home. A steam winch amidships was used as the motive power, coupled to two wooden spars running in wooden plummer blocks lined with sheet metal, and these spars were allowed to protrude through holes cut in the bulwarks. Wooden arms with planks nailed across them formed the paddle-wheels and with this contraption the ship made 1½ knots and eventually reached Hampton Roads in safety.

The most famous paddle steamers were, of course, the North Atlantic passenger vessels, the ocean greyhounds of the age, starting with the *Sirius*, the ship that made history by being the first ever to cross the Atlantic by steam alone. She was a smallish ship, of 412 tons, built in 1837 at Leith, for the St George Company with side-lever engines by J. Wingate & Co. of Glasgow, there being two cylinders 60 inches diameter by 6 feet stroke, giving her a speed of 8½ knots with a rectangular flue boiler working at 5 lb. Appropriately, her figurehead was that of a dog rampant. She sailed from London on 28th March 1838, and called at Cork to pick up passengers and mail, leaving on 4th April for the States. Her log shows the usual entries for storms and bad weather normal at that time of the year, and the chief engineer's habitual preoccupation with fuel for the voyage, commonly referred to as coal fever. In this case, rosin was mixed with selected ashes to eke out the coal and, incidentally, clean the funnel. She also made several stops to repack the stuffing boxes, also to tighten the screw on the eccentric and once to refasten the floats. On several occasions the fuel consumption was checked and worked out at 1 ton coal per seventy-five minutes for 320 nominal horsepower, the total coal consumed being 450 tons.

The *Sirius* made several voyages across the Atlantic before she settled down to coastal trade, running between English ports and Southern Ireland. She stranded and broke up on the rocks at Ballycotton Bay in 1847 and made history of a sort in her passing, much as she did on her maiden voyage. Her cargo consisted mainly of barrels of porter and these were being washed ashore onto the Ballycotton beaches to the everlasting joy of the local residents. A force of 200 marines were hurriedly despatched to the spot from Cork, but they arrived too late to save the cargo and stayed only long enough to bury two villagers, who, having drunk themselves to death, had achieved their own special state of Hibernian Nirvana. A further historical note here is that the main shaft of the *Sirius* was salvaged and built into the engine of a mill outside Cork and was in use until 1899.

Another famous vessel was the *Great Western*, built at Bristol in 1837, which shared with the *Sirius* the honour of the first Atlantic crossing, being beaten by a margin of hours only. This vessel was considerably larger than her competitor, being some 1340 tons, with engines of 750 nominal horsepower and a speed of 11 knots at fifteen revolutions per minute of her paddles. She had side-lever engines with twin cylinders 73 inches diameter and 7 feet stroke, paddle-wheels 28 feet in diameter by 10 feet width of float, taking steam from four flue boilers at 6 lb pressure. These boilers were subsequently renewed by tubular boilers at 12 lb pressure, and both ship and machinery, designed by that Victorian genius Brunel, proved very successful for the whole of their twenty years' life-span.

The largest purely paddle steamers on the Atlantic run were the Cunarders *Persia*, *Arabia* and *Scotia*: the *Persia* being the largest in the world at the time of her launching, the *Arabia* being one of the first steamers to have steam heating in the accommodation, and the *Scotia* being the last of the great paddle steamers on the Atlantic service. One other vessel may be mentioned, the *Great Eastern*, the largest vessel of the nineteenth century, but she was driven by both paddles and screw and a description of her machinery is reserved for a later chapter.

5 Screw engines

Although the paddle steamer dominated the national scene to the almost total exclusion of the screw steamer during the first half of the nineteenth century and well into the second half, the steam-driven screw propeller and the paddle-wheel appeared simultaneously. Colonel John Stevens in America designed a screw-driven steamer in 1804 before he built his paddle steamer *Phoenix*, and experiments on screw propulsion had been made long before that date.

The earliest reliable record of a successful screw propeller is contained in a certificate issued by Admiral Sir Richard Bickerton dated 4th July 1802 and endorsed by the captains of several naval vessels stationed at Gibraltar, who witnessed a demonstration of Shorter's propeller applied to the *Doncaster*. This was a device with spokes and floats attached to the vessel's stern with the after end kept afloat by a buoy. The wheel was rotated by a horizontal shaft driven by ropes from the capstan by eight men, and the certificate claims the vessel made slightly more than a knot in a flat calm. It is possible that the same number of men in a pair of boats would have towed the *Doncaster* at the same speed, but the experiment does show that the principle of the screw had been discovered and tried out.

John Stevens and John Ericsson are generally acknowledged as the joint inventors or, properly, the joint developers of the screw propeller, although many others have advanced the basic design, and every early screw-driven ship had a different-shaped propeller. However, in spite of the weight of invention behind it, the helical propeller took some fifty years to gain recognition: there was nothing wrong with the screw as such, there was just no type of machinery to drive it. The only steam engine available was the ponderous slow-speed paddle engine with its huge cylinders and heavy running gear, not at all suitable to drive a screw propeller which, like the petrol engine of a later date, has a certain minimum speed below which it cannot deliver any useful work.

Screw propulsion posed several problems to the Victorian marine engineers which they had not encountered in the paddle steamers. The main one, of course, was rotational speed and this was impossible at the then state of engineering knowledge and the low boiler pressures of the time. High speed demanded high pressure, and this came into general use only in the 1860–70's. The alternative was to use a heavy slow-running engine geared up to the propeller shaft, and while this system was often used, the gearing was found to be bulky, noisy and expensive. Gear drive was used in the early screw steamers for precisely the opposite reasons as for the early paddle steamers. In the latter the engine was light and had to drive a pair of heavy paddle-wheels rotating in a fairly 'solid' medium, water. Gearing was necessary to reduce engine to paddle-wheel speed. In the former case, gearing was again necessary to increase the relatively slow-running engine to the higher shaft speed.

Another problem was that of absorbing the propeller thrust into the ship's fabric, a problem only recognised after many screw installations had been completed. In the paddle drive there was no problem, as the forward thrust of the wheels was taken up by both paddle and intermediate shafts and transmitted to the framework through their plummer blocks which were specially adapted for this purpose and were often inclined to the horizontal to present more of the bearing crown to the line of pressure. The early screwshaft driven by gearing ran in ordinary plummer blocks and the forward end of the shaft was fashioned with a wide flange that bore against a corresponding gunmetal flange fixed to a thrust post on the keelson. The propeller thrust was thus absorbed directly by this post from the shaft itself, the steel-to-gunmetal rubbing surfaces being kept lubricated and cool by a copious sea-water spray.

When the propeller was direct coupled the thrust was transmitted to the crankshaft, and this had to spread the load throughout the engine bedplate by taking the strain on the cheeks of the bearings, a most unsatisfactory arrangement indeed, and one that caused considerable trouble and overheating until the horseshoe thrust carriage came to be used.

Perhaps the most difficult problem to solve was that of finding suitable means of providing an underwater fitting in the ship's hull for the main shaft; one, moreover, that would carry a very heavy overhanging load. A fitting, in fact, that would act as a bearing, a steady and stuffing box all in one. The problem was tackled by fitting a good strong cast-iron stern tube through the sternpost with a stuffing box on the inside of the vessel, and the screwshaft running directly on the iron surface with sea-water lubrication. This caused very heavy wear entailing frequent renewal of both shaft and tube, and after some experimenting the stern tube was bushed with gunmetal, making for a quicker and cheaper repair. The stern arrangements, however, remained a headache for both marine engineer and owner until John Penn invented and patented the lignum vitae lined stern bush in 1854.

Finally, and here we are on rather delicate ground, the primary reason for keeping to paddle drive instead of changing to screw propulsion lay directly with the Admiralty, who insisted that paddle-wheels should be a condition of the mail subsidy. The first screw-driven ship in the P. & O. Company was launched in 1853 and in the Cunard as late as 1862: no new-fangled foreign contraption on the mail ships in those days.

The first steamer to be propelled by screw was the *Archimedes*, a small wooden

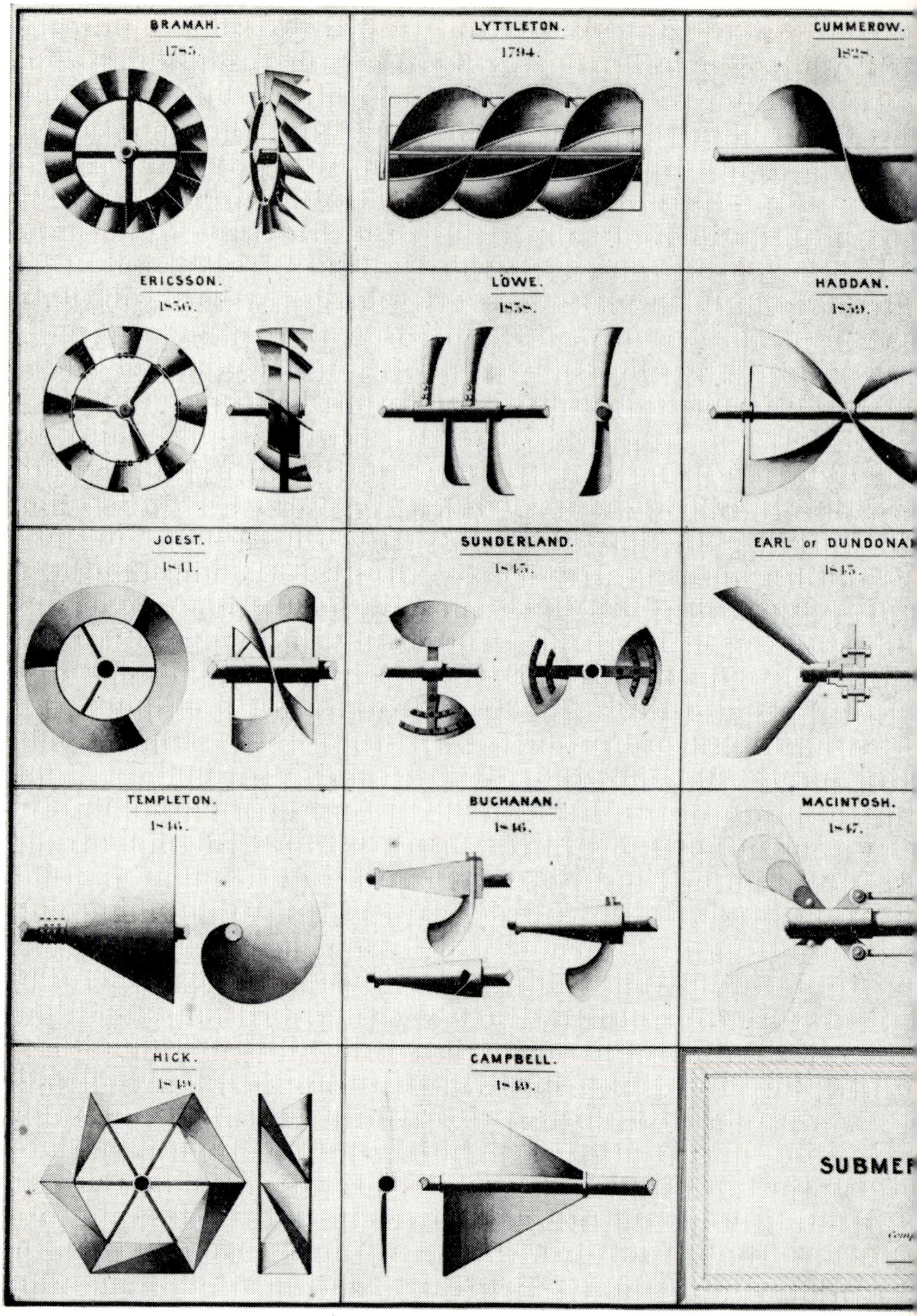

21 Various types of early screw propellers

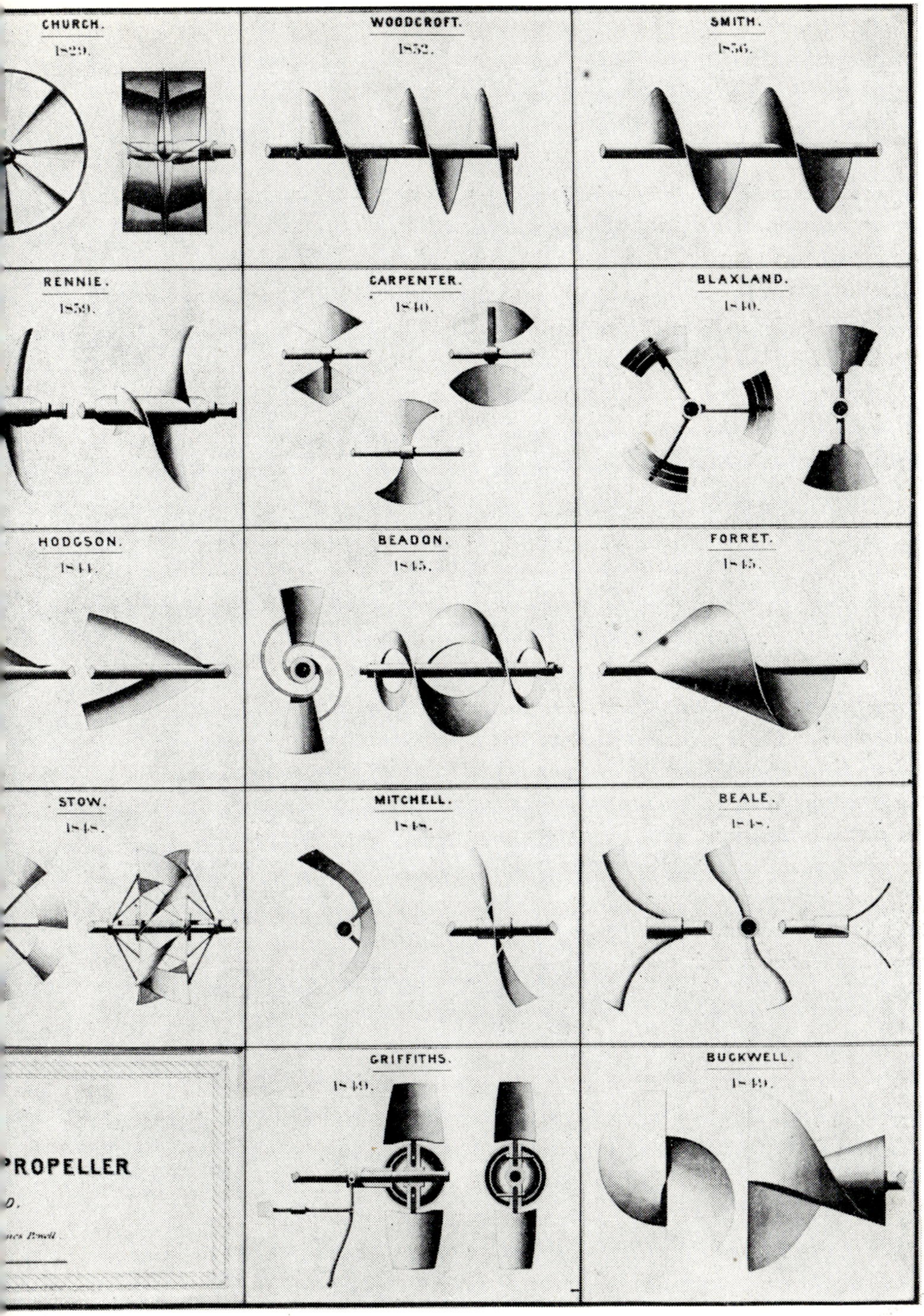
CHURCH.
1829.
WOODCROFT.
1832.
SMITH.
1836.
RENNIE.
1839.
CARPENTER.
1840.
BLAXLAND.
1840.
HODGSON.
1844.
BEADON.
1845.
FORRET.
1845.
STOW.
1848.
MITCHELL.
1848.
BEALE.
1848.
PROPELLER
GRIFFITHS.
1849.
BUCKWELL.
1849.

schooner of some 240 tons built in 1838 by Henry Wimshurst and engined by J. and G. Rennie, using Francis Pettit Smith's patent propeller. The engine had twin cylinders 37 inches diameter by 3 feet stroke, and developed 80 NHP at 26 rpm on a boiler pressure of 6 lb. The screwshaft was geared up to 140 rpm and the propeller itself consisted of a single complete helix 7 feet diameter and about 8 feet pitch, giving a ship's speed of 8 knots.

There is no doubt that Brunel was very much influenced in his design of the *Great Britain* by the performance of the *Archimedes*, especially after she had completed a voyage round the British Isles and to the near Continent, but apart from her fame as pioneer in screw propulsion, the *Archimedes* did not cut too good a figure in nautical history; she had a boiler explosion shortly after her maiden voyage and broke her crankshaft after only one year in service. To crown this achievement she went ashore in 1840.

By 1843 there were already about a dozen screw ships in service, including the *Great Britain*, the *Rattler*, *Princess Royal* and other famous vessels, and by 1850 the screw steamer had become established.

Twin-screw steamers arrived a little later on the scene, in the 1860's, and they had to wait until shipbuilders and engineers had mastered the techniques of underwater fittings and high-speed engines. In the early screw ship it was difficult enough to site a stern tube at the stern frame, probably the strongest member of any ship, without having to worry about a pair of underwater bearings on the quarters with no special framework to build on to. Also, even with a single screwshaft the engine room was embarrassingly full of hardware without having to find room for a second engine.

The early iron-cased battery ships had been driven at times by twin screws, but both shafts had been geared together and worked by a single engine, and the first vessels to be propelled by twin screws independently driven were the *Flora* and the *Far East*. The *Flora*, built in 1862, was a steamer of 400 tons fitted with a a pair of compound annular engines developing a total of 120 NHP, with cylinders 26 inches diameter and 21 inches stroke taking steam at 30 lb. She was unusual in that she had two main boilers working at 30–35 lb and an auxiliary boiler providing steam at 50 lb just to produce a steam blast at the base of the funnel to cause an induced draught and provide a certain amount of dryness in the main steam line. Another feature, uncommon in normal steam engines but not unusual in annular engines, was the three piston-rod arrangement whereby the two rods from the LP annular piston and the single rod from the HP central piston were all coupled to the one crosshead bar.

The *Far East* was a much larger vessel of some 2200 tons displacement, with a pair of twin-bladed propellers housed in lifting frames, each of 8 feet 2 inches diameter by 16 feet pitch, and she made her appearance at about the same time as the *Flora*.

Multiple-screw drive became inevitable with the increase in size of the ships and their engines, and just as the power increased, so did the propeller size increase to absorb this power. Towards the end of the century the mere question of propeller size became a problem: in ships of 20,000 tons and over, with engines developing some 10,000 HP per shaft at fairly low revolutions, the disc diameter attained huge proportions and in twin-screw vessels the deadwood aft had to be cut away to allow the propellers to work in this gap. There came a point where the port and starboard propellers were so large that their discs overlapped and this involved stepping one propeller ahead of the other to clear the blades.

This solution was not very satisfactory in that one propeller caused interference and cavitation in the other, but it was a compromise necessary in a fast passenger vessel with powerful engines and a fine run. In the case of the Norddeutscher Lloyd steamer *Bremen*, this type of propeller overlap very nearly caused foundering on top of disaster. In 1905 her port screwshaft broke inside the stern tube and the propeller gradually slipped back until it fouled the blades of the starboard propeller and locked with them. The vessel was lucky in being discovered and towed into Halifax with both screws broken.

Reverting to the early single-screw ships and the mail subsidy, it was the Lords of the Admiralty themselves who demonstrated the superiority of the screw-propeller drive to their own and nearly everybody else's satisfaction. In April 1845 they arranged a tug of war in Yarmouth Roads between two sloops: the screw-driven *Rattler* against the paddle-wheel-driven *Alecto*, both vessels being identical as regards displacement and horsepower. Subsequently, in 1849, another back-to-back trial took place between the screw corvette *Niger* and the paddle sloop *Basilisk*, and on both occasions the screw-driven vessel towed the other backwards. On the second test the displacement and horsepower were greater for each ship than in the case of the earlier trial, so that scale effect could be discounted, but it was claimed at the time that with each vessel at a standstill straining one against the other, the propeller would act as a centrifugal pump and generate a considerable wave on the *Rattler*'s stern which would tend to move it ahead.

As both propeller and paddle-wheel work on the same principle in somewhat different ways, perhaps the best comparison of two such identical vessels would have been that suggested by a merchant ship-owner: by relating speed to consumption over a determined distance. When all is said and done, the efficiency of the two modes of propulsion is invariably shown in the accountant's balance sheet without having to pit one ship against another in an expensive and rather inconclusive experiment.

Just as in the case of the paddle steamers with their disengaging wheels when proceeding under canvas, so it was with screw propulsion: some means had to be found to do away with propeller drag when under sail with the engines stopped and here it might be pointed out that steamships never paid their way until the arrival of compound expansion, in the early 1860's. Until that time, with boiler pressures of the order of 20 lb and fuel consumption at, say, $4\frac{1}{4}$ lb per IHP per hour, engines were a luxury and the ship-owner never lost an opportunity of pointing this out to the master. As soon as a favourable wind sprang up, canvas was set and fires were extinguished. Apart from sheer cost, however, other considerations had to be taken into account. Away from the large ports, there were few bunkering stations about the world, and any vessel that ran short of fuel had to revert to a sailing ship. This entailed no special hardship, as most vessels throughout the nineteenth century were rigged for sail whether they used them or not, and all hands, including engineers and firemen, were quite used to handling ropes.

There were three different ways of reducing drag in the screw steamer: by disengaging the screw shaft, by feathering the propeller blades or by lifting the propeller clear of the water. The first was quite easy to achieve; a dog clutch somewhere along the line of shafting could be thrown out and the propeller allowed to trail. The second method, used by Joseph Maudslay, involved a propeller with two movable blades operated by a control rod located within the screwshaft and connected to a sliding collar fitted on the forward end of the shaft. This collar was moved by means of a vertical spindle from the deck and in turn it rotated the propeller blades about their vertical axes.

The third method, that of lifting the propeller out of the water, was by far the most popular and was still being used for ships building as late in the century as the 1870's. The two-bladed propeller, with a short stub shaft, was set in a pair of bearings in a banjo frame fitted into the stern-frame aperture in such a fashion that the whole frame could be lifted into a trunk in the counter by means of pulleys or screws, according to size and weight. The forward end of the stub shaft was made with a projection to slide into a corresponding slot in the after end of the screwshaft, like a dog clutch, and the two stub-shaft bearings rested on brackets on the banjo frame forward and aft of the propeller, there being dowels to position the frame correctly. When the engines were running, this banjo frame was tommed into position by a wooden shore or by the operating screws. The forward and after sides of the stern-frame slides were provided with safety racks to engage a pair of pawls on the frame itself. When lowering the frame, a line was attached to each pawl and kept tight until the propeller was safely in position on the supporting brackets.

To operate the lifting gear, the engine was barred round until the propeller blades were in the vertical position to allow the locking gear on the frame to engage the tip of the blade, to prevent swinging. The shore was then removed and the frame raised by hand, the pawls dropping into the rack teeth as the frame ascended. When clear, the frame was secured by a locking device to prevent damage, a necessary precaution in view of the weight of frame, propeller and shaft.

The banjo frame was not the only type of gear for lifting the propeller clear when under sail, there were a few other types all more or less elaborate, and possibly the oddest was that designed for the Oceanic Steam Navigation Company (later White Star Line) steamer *Britannic*, launched in 1874, and the fastest steamer of her time. This was a four-masted barque-rigged passenger and cargo ship built in Belfast some 455 feet long by 45·2 feet breadth and 33·7 feet depth, registered at 5000 tons. Her two-crank compound engine with tandem cylinders of 48 and 83 inches diameter by 5 feet stroke developed 5000 HP and drove a propeller 23 feet 6 inches diameter by 31 feet pitch at about 50 rpm. This monster propeller was hoisted up into a trunk in the counter in a startlingly novel way. The stern frame was made to include a very high aperture, while the propeller post was forked to allow the propeller and screwshaft in its stern tube to be lifted up into the trunk at the top of the stern-frame arch. The last section of intermediate shafting passed through a stuffing box on the after peak bulkhead and was supported on a very heavy bearing inside the peak tank. A universal coupling at the after end of this intermediate shaft engaged with a similar coupling on the forward end of the screwshaft, and acted as a fulcrum when the propeller was swung up.

The operating mechanism for lifting this massive gear was situated under the poop deck, a hand-worked capstan driving a worm and wheel on the lower deck, which, through rack and pinion, lifted the iron links attached to the stern tube. When one considers the size of the vessel and her machinery, and of the weight and complexity of the stern arrangements, one might be excused for wondering if it was all worth while. This lifting gear was soon found awkward to handle and the propeller shafting was modified to drive in the usual way.

Although the banjo lifting frame was easy to operate, it was not foolproof, as witness the damage to the steamer *Croesus* in 1853. This vessel on a voyage from UK to Australia arrived off the Cape of Good Hope under sail and proceeded to lower her propeller into position before entering Cape Town harbour under steam. Unfortunately, the lowering tackle parted and the banjo frame slid down

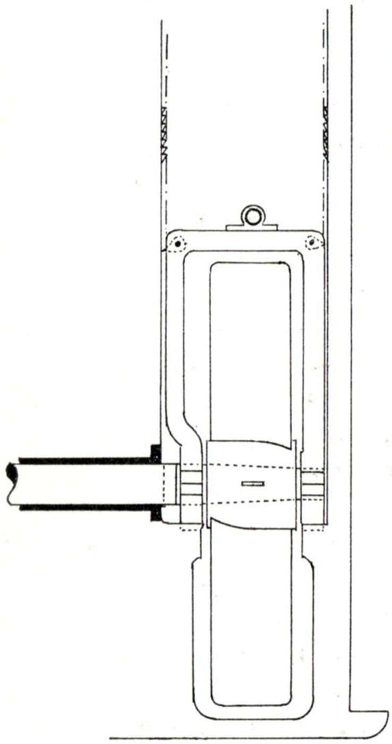

22 *Left:* Banjo frame for lifting screw propeller when under sail
23 *Right:* General view at platform level of a horizontal screw engine (1870)

with a run, breaking the stern-frame sole piece and starting the riveting in the after peak tank and bulkhead boundary bar. The vessel was towed into Cape Town and the bulkhead made tight with a cement box of clay and clinker, while the lifting frame was jacked up and reset on its bearings. A permanent repair was carried out at Sydney in a most ingenious and novel manner for that age: a cofferdam was built around the rudder and stern, extending forward of the bulkhead and made tight against the hull with greased sheepskins, and with a small well at the bottom for the hand pump. The sole piece was repaired and any loose riveting made good, although it took the repairers four months to make and fit the cofferdam, and once it was fitted, pumping had to be continued day and night to keep it in place.

It may be recalled that the *Great Eastern* had an extensive hull repair carried out in New York in 1862 using a similar cofferdam attached to the hull, and that historians of the day claimed this method of underwater repair as the first instance of its kind.

Whilst on the subject of damaged stern gear it might not be out of place to mention a smart piece of dry-dock work performed in mid-ocean in the days before wireless telegraphy was dreamt of. The *Cereda*, on passage Tenerife to Pensacola, broke her screwshaft halfway across the Atlantic, off the steamer route. The ship was trimmed heavily by the head to bring the stern tube above water and allow it to be plugged. The broken piece of the screwshaft was thereupon removed and replaced by the spare screwshaft, and the wooden plug knocked out. Meanwhile, the spare propeller had been hoisted up from the hold and fleeted aft on the derricks, while a spar had been rigged across the stern to support it in place when inserting the screwshaft cone. With the rolling of the ship and the banging of the spare propeller against the new shaft, the threads were badly damaged and the engineers had to be swung down on bosun's chairs in relays to dress them. With the best will in the world, the nut could not be flogged more than about halfway up the thread, but it was sufficient to allow the ship to steam ahead and arrive at her port of destination only twelve days late.

The screw steamer, coming on to the scene at a time when the paddle engine had been developed and refined into a reliable and efficient prime mover, was saved a lot of experimenting and could adopt without too many changes a number of the paddle-engine types which had proved themselves on the high seas. It still had two very tough problems to solve, however—the screwshaft had to run at about four times the speed of a paddle shaft and, in the case of a direct drive, the crankshaft had to be placed fore and aft on the centre line of the ship.

The first problem was tackled in the same way as in the pioneer paddle ships, by gearing. The crankshaft carried a large wheel made up of three or four separate castings all clamped together and fitted with wooden cogs in their periphery all of the same size and pitch, the teeth of one wheel being slightly in advance of those in the neighbouring wheel to ensure continuity of engagement. The driven pinion on the screwshaft was a solid casting with teeth moulded in. This form of gearing was bulky, expensive and very, very noisy. It was, moreover, a perpetual and horrible menace to the engine-room crew when in motion.

The second problem was never satisfactorily solved until marine engineers stood back and made a reappraisal of the whole principle of screw propulsion. In naval vessels it had been the practice to place the steam cylinders as low as possible in the engine room, preferably right on the bottom, but at least below the water line out of range of enemy shot. In merchant ships there was no need for this precaution and in the slow-running side-lever engine the vertical cylinder placed on top of the keelson was probably the best place for it. But when it came to direct-drive screw propulsion, marine engineers were reluctant for some reason to abandon the horizontal position for their engines.

Until the 1870–80's steamers still tended to follow sailing-ship practice in their cross-section with a generous round of bilge and a fairly steep rise of floor. This gave very little space on either side of the ship's centre line to place an engine and several designs were brought out to cope with the situation. Some builders used the return connecting rod (steeple) type with opposed cylinders while others introduced engines with a novel form of connecting-rod attachment, or dispensed with the piston rod in an attempt to shorten the engine.

The horizontal oscillating engine was fairly popular in the early direct-drive screw engines, there being either one or two cylinders placed on each side of the crankshaft with the piston rods laying directly on to the crankpin.

Another popular type was the steeple engine, again with two or four cylinders, these being placed on either side of the crankshaft facing each other with their piston rods connected together on to a common crosshead. A connecting rod coupled to this crosshead drove on to the crankpin. In this type of engine the cylinder arrangement in plan was made lopsided to give length to the connecting rod. The cylinders on one side of the shaft were moved outboard a little to give reasonable movement for the crosshead, while the opposite cylinders were as close as possible to the crankshaft.

The drawback to the two types of engines described when four cylinders were fitted was lack of accessibility. These engines invariably had their air pumps between the two sets of steam cylinders and their running gear had to be felt round and lubricated. This meant clambering over smooth, slippery, oily surfaces to reach the moving parts, and marine engines in the early Victorian years were remarkably free of guard rails and safety devices which cost money but paid no dividend.

The Gorgon-type horizontal engine was also used, although it never achieved much popularity until it was redesigned and made vertical inverted, to become the standard marine steam engine for the rest of the nineteenth century and well into the twentieth.

A type of horizontal screw engine that was never acknowledged on this side of the Atlantic was Ericsson's divided cylinder with bellcrank drive which was used in the early *Monitor* battery ships during the American Civil War. It consisted of a long cylinder with a central diaphragm to form two steam cylinders

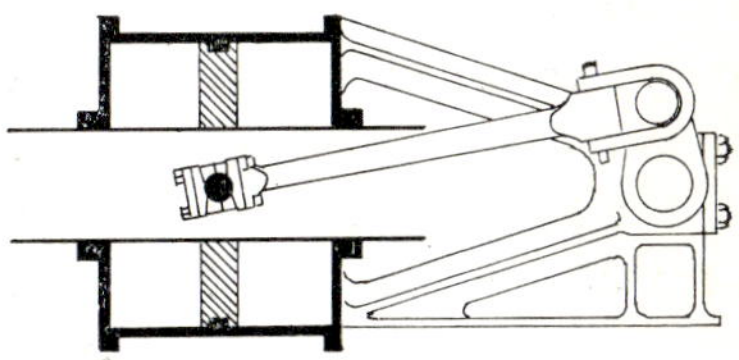

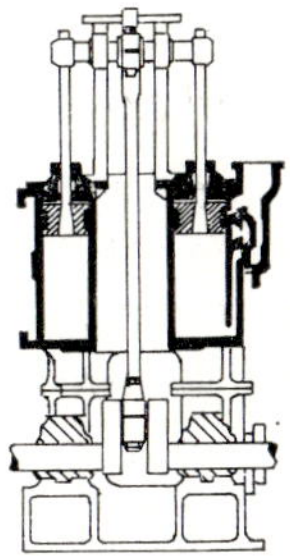

24 *Left:* Ericsson's divided-cylinder screw engine with double bellcrank lever drive (1860)
25 *Right:* Early screw engine types:
(*above*) Penn's trunk engine
(*below*) Maudslay's annular engine

placed bottom to bottom, each with its own piston, valve chest and gear. Each piston worked on to a separate bellcrank which in turn drove a common crankpin through a connecting rod, there being separate arms on both lay shafts to drive the air pumps. The beauty of this arrangement lay in the fact that owing to the angularity of the bellcranks and their connecting rods, it was impossible for the engine to stall on centres.

Two entirely new engines made their appearance in the 1850's, of a type completely unknown to builders of paddle steamers: the trunk engine and the annular engine. These two types have always been confused, the one with the other, possibly because both are in fact trunk engines and both are annular engines, although in each case working on an entirely different principle. James Watt introduced a form of trunk piston into his wooden steam hammer patented in 1784 but it was John Penn who established the system in the marine field. The engine was essentially a horizontal one and the piston was cast with a large hollow pipe or trunk passing through the centres of both the front and back cylinder covers, there being a large stuffing box at each end. A gudgeon pin was bolted to the centre of this trunk, in line with the piston, and a connecting rod coupled to this gudgeon pin drove the crankpin direct. The trunk was made large enough in the bore to take the swing of the connecting rod without fouling, and in the case of a right-handed propeller, the steam cylinder would be placed on the starboard side of the shaft so that in the ahead running condition, the connecting-rod reaction would tend to lift the piston and so counteract the weight of the moving parts. In this engine the air pump was placed opposite the steam cylinder and on the other side of the shaft, the drive being taken from an auxiliary piston rod screwed into the piston and emerging through a stuffing box on the front cover. This meant that the air-pump stroke was equal to the main piston stroke and to cut down excessive waste of power the pump was made of fairly small bore and double-acting, i.e. it had a solid bucket, the foot and head water valves being in separate cages at each end of the cylinder.

This engine was successful possibly more on account of its workmanship and detail than for its leading features. The trunk was found to give trouble at the glands due to ash and dust—in those days the boilers were still placed in the engine room and firemen then, as now, were still as happy-go-lucky in dowsing the ashes when cleaning fires, the major cause of dust and soot in any engine room. Apart from that, however, the exposed trunk meant considerable heat loss by radiation and condensation in the cylinder.

By contrast, the annular engine was in the main a vertical one, with the cylinder placed just sufficiently high above the crankshaft to clear the webs and so remain below the water line, thus qualifying for acceptance in both naval and merchant ships. The cylinder shell was cast in one piece with a central trunk, there being separate top and bottom covers. The piston was made in the form of a large doughnut with piston rings bearing on the outer cylinder walls and the inner trunk walls, and a pair of piston rods extended above the cylinder to a crossbar running on guides attached to the top of the trunkway. From this high crosshead a long connecting rod passed through the centre of the cylinder trunk to lay on to the crank below, the trunk being, of course, wide enough to clear the swing.

This design had certain features in common with the return-connecting-rod or steeple engine but it also shared the disadvantages of the trunk engine in its high steam consumption.

The most successful of all direct-acting screw engines was that originated by J. and G. Thompson, the so-called steamhammer or inverted engine: a vertical Gorgon engine having the steam cylinders placed well above the crankshaft with

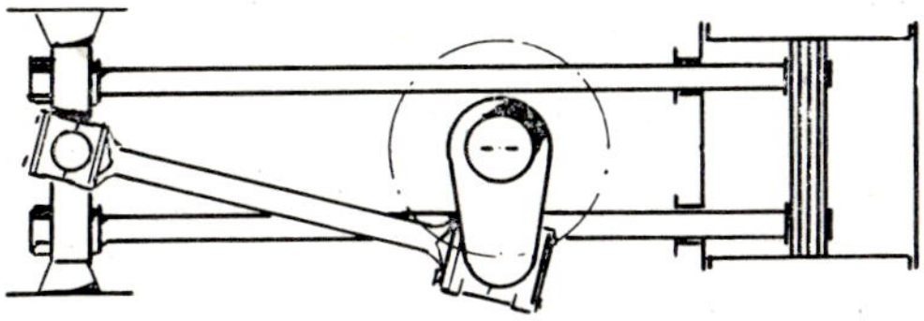

26 Horizontal return connecting rod screw engine

the piston rod and connecting rod in between. Although slow to attract attention, and out of the question for unarmoured naval vessels, it eventually came to be acknowledged as the only form of engine for screw propulsion and has survived a century of experimentation with many engines working on widely different principles.

Screw steamers did not make the same impact on the public imagination as did the early paddle boats: there was a vast difference between crossing the Atlantic by sail, more or less at the mercy of the elements, and making the same crossing under steam. Both voyages meant acute seasickness, stuffy cabins, poor food and sheer misery to the poor passenger, but he suffered much longer in a sailing ship. As soon as steamers became rather more than experimental, they inevitably took over the passenger trade, and by 1850 there was never any doubt in the passenger's mind as to how he should make the crossing. But, as to the difference between screw steamer and paddle steamer, that was a very minor point to the tourist trade, comparable to the turbine and oil-engine drive in our present-day passenger steamers where few other than the crew are aware of the difference. By the 1860's there were many equally large, fast and luxurious liners driven either by paddle or screw.

Of the many thousands of screw steamers built in the nineteenth century many were famous after a fashion but two stand head and shoulders above the crowd: the *Great Eastern* by reason of her size—she was the largest ship (and the biggest white elephant) of the century, and the *Great Britain*, partly on account of her peculiar build but mainly because of her remarkable history which is not yet complete. By a coincidence (or is it?) both vessels were designed by I. K. Brunel, that acknowledged master engineer of his time. The *Great Eastern* had both paddles and screw propulsion and the details of her machinery are reserved for a later chapter, but as the *Great Britain* has recently come into the news in connection with her long voyage home to the UK and set her up as a museum, a brief outline of her career and description of her original machinery might not be out of place.

The Atlantic Ferry was inaugurated by the Great Western Steamship Company in 1837 with their new wood paddle steamer *Great Western*, and the service proved to be so lucrative that the Company decided to order a second, larger ship to be built entirely of iron, designed by their consultant Brunel. This new vessel would be the largest in the world but as nobody had had any experience of building iron steamers approaching the size contemplated, the company were forced to build both ship and machinery themselves. A section of the dock adjoining Cumberland Basin in Bristol was allocated to the Company as a building berth and the keel was accordingly laid in 1839, the entrance to the berth being blanked off during construction and the dock pumped dry.

Now it had been understood by the builders that improvements to the docks were shortly to be undertaken by the Bristol port authorities, including, among others, the widening of the existing locks in the Cumberland Basin and its approaches, and in anticipation of this, the beam of the new ship was made some 5 feet greater than the lock width. Furthermore, it had been assumed that the lock walls were reasonably perpendicular, but when the shipbuilders came to survey the underwater portion they found the walls sloping outwards towards each other. This was rather a setback as the effective width of the lock, even after widening, would fall far short of the nominal width. The *Great Britain* was therefore made with an unusual trapezoidal section, the width at bottom being only half the extreme width over belting, there being an excessive tumble-home to meet the weather deck. This peculiar section was considered necessary to take full advantage of the shape of the locks and still retain her designed capacity, and the resultant lines of the hull were deemed to be not only very shipshape to the critical nautical eye but most pleasing to the aesthetic sense; this ship did not have a flat plate in her.

In the event, all this inventive energy spent in deciding the shape of the hull was in vain, the promised dock improvements did not materialise and the basin locks remained as they were. The shipbuilders, having finished the ship, were now in the position of the man who built a yacht in the spare bedroom. After months of argument and negotiation, the Great Western Steamship Company were allowed to widen the lock entrances to allow the *Great Britain* to pass through, and then to restore the locks to their original dimensions. All this at their own expense, a Cesarean operation on a monumental scale.

The official launch, at Bristol, took place in 1843 and her trials were carried out in 1845, mainly on the passage round to London, six years after the keel was laid.

Like all good Victorian ships she was designed primarily as a sailing ship with six masts, the second or main mast only carrying a square mainsail and a top-

27 SS *Great Britain* (1843)—Arrangement of diagonal house engine with chain-driven screw shaft

sail, the other five being rigged with spanker and gaff topsail, the foremast in addition being equipped with a staysail, jib and topmast staysail; all in all a difficult rig to categorise. The mainmast was a built mast stepped in the ordinary way on to the keelson, but the others, the fore and aft rigged masts, were stepped into wrought-iron tulips fitted on the weather deck.

It had been Brunel's original intention to propel the *Great Britain* by paddles, but in view of the success of the screw steamer *Archimedes* in 1838, he arranged for the Great Western Steamship Company to borrow the *Archimedes* for the purpose of trying out the performance of various types of propeller.

The propelling machinery for the *Great Britain* consisted of a four-cylinder diagonal direct-acting engine, there being a pair of cylinders working on to an overhung crankpin at each end of the shaft with the main driving wheel between them. The cylinders, inclined towards each other, were 88 inches bore by 6 feet stroke, and had their jet condenser placed between them. The engine was of the 'house engine' type, i.e. an engine without a bedplate but built into the fabric of the ship, in this case the framework consisting of a pair of hardwood trestles or A frames, each supporting one pair of cylinders. The crankshaft sat at the apex of the trestle arrangement with the intermediate shaft below it. The air pumps, placed between the A-frame legs, were driven by the main crankpins, while the feed and bilge rams were operated by the air-pump parallel motion.

The valve drive was of the single eccentric type, with a separate expansion valve gear. The main eccentric was mounted on a large toothed wheel which, when reversing, could be partly turned by a hand steering wheel and pinion to bring the eccentric sheave into the astern position, care being taken, of course, to withdraw the pinion before the engine commenced turning under steam. The crankshaft was a heavy forging 17 feet long and weighing 16 tons in the rough, and was 30 inches diameter at the journals. A hole was bored through the shaft, the webs and the crankpins, into which a stream of cold sea water was forced at boiler feed pressure, and this served to cool the journals and lubricate the bearings. As the crankshaft was situated at the top of the engine, this watercooling system must have resulted in a messy engine room, and an unpleasant business for the watch engineers feeling round. Between the two pairs of cylinders was placed the main gearwheel, a large toothed drum 18 feet 3 inches pitch diameter and 38 inches wide, keyed to the crankshaft. This wheel was made of iron very much in the style of a paddle-wheel, and it drove a 6 feet diameter pinion keyed to the propeller shaft immediately below, through a heavy chain-link belt, the cogs in the wheel being of teak in the large wheel, and of lignum vitae in the pinion, while the chain links were made of wrought iron.

The intermediate shafting consisted of three sections, the forward one, a solid forging 16 inches diameter and 28 feet long, and the after section 18 inches diameter by 25 feet long. The middle length was a hollow cylinder 62 feet long and 30 inches in diameter fabricated from $\frac{3}{4}$-inch plating riveted together. The propeller thrust was taken up by a 24-inch-diameter gunmetal disc on the forward end of the shaft, bearing against a similar steel disc built into the engine framing.

The propeller was of wrought iron with six spokes or arms, on the ends of which were riveted the iron blades 4 feet 3 inches long on their outer edges by 2 feet 10 inches deep, and $\frac{7}{8}$ inches thick. This propeller was, in fact, more of a paddle than a propeller as we know it. Its diameter was 15 feet 6 inches at 25 feet pitch, and at full speed it turned at 18 rpm.

Like everything else about this ship, the boiler was unusual. The common casing was 34 feet long by 31 feet broad, and stood 21 feet high and it contained three separate boiler systems in it, each with its own four furnaces forward and aft and with separate flues, joining together at the base of the funnel. There was also a feed-collector tank built round the lower part of the funnel taking delivery from the engine-driven feed pumps and from this feed heater the water flowed by gravity into the main boilers, the working pressure being kept at 5 lb. To complete the overall picture of this early steamer, her vital statistics were 322 feet length overall, 50 feet 6 inches extreme breadth, 32 feet 6 inches with a loaded displacement of 3620 tons.

The *Great Britain*'s maiden voyage to New York was uneventful, the average speed working out at 9 knots although she had logged 12¾ knots on trials. The second voyage was not so happy; after only five days out from Liverpool her foremast went over the side in a squall and fouled her propeller, breaking off numbers 1, 4 and 5 blades. The remaining blades, with the aid of her sails, managed to get the ship to New York where, after discharge, she was placed on a hydraulic grid for repairs, there being no graving dock in New York harbour at that time, and the propeller was equalised by removing blade number 3 and riveting it into the place of number 4.

This hydraulic grid was a curiosity in itself, and consisted of a large gridiron with keel blocks along its centreline, sunk between two jetties spaced a ship's width apart. A huge hydraulic ram at the shore end of each jetty carried a yoke to which were attached a number of strong chains of different lengths laid along the jetty and guided on rollers. At intervals these chains ran round sheaves and on to a cross-member of the gridiron. With the vessel placed over this framework and centralised above the keel blocks, the ram pumps were set going and the gridiron was hoisted above water level.

With her propeller balanced again, albeit with only three blades, the *Great Britain* set off on the homeward passage, but after only a few days at sea she shed the remaining blades and completed the trip under canvas making 10 knots with her engines stopped.

Upon arrival at Liverpool a new four-bladed cast-iron propeller was fitted instead of the original six-bladed wrought-iron one, and the vessel resumed her Atlantic ferry service without further incident until the end of 1846 when she stranded in Dundrum Bay, Northern Ireland, in a storm on passage to New York. There she remained for eleven months battered by wind and tide, having been written off by the underwriters, until she was floated off and towed back to Liverpool for survey on the gridiron. She was subsequently sold to Messrs

Gibbs, Bright & Co. of that port and completely overhauled. Her six masts were removed and replaced by four stepped in the orthodox manner, with a spread of canvas equal to the original sail plan. Her machinery was gutted and replaced by a pair of geared oscillating engines supplied by John Penn & Sons of Greenwich with cylinders 82½ inches bore by 6 feet stroke, driving a three-bladed cast-iron lifting propeller 15 feet 6 inches diameter by 19 feet pitch. The screwshaft gearing was renewed by wheel and pinion of 14 feet and 4 feet 6 inches diameter respectively, the ratio remaining at 1:3, and the old single-casing boiler arrangement was pulled out and replaced by six tubular box-shaped boilers working at 14 lb. The original engine horsepower of 1000 was by this means increased to 1450 and the sailing speed increased to 12 knots.

The rebuilt *Great Britain* was placed on the Australian trade in 1853, carrying passengers and convicts, and became very popular on that run, with the paying passengers that is, and remained in it for twenty-five years. After this she underwent a further conversion, had her machinery removed entirely and became a full-rigged sailing ship on the bulk-cargo trade. Finally, in 1886 she sustained severe damage in rounding Cape Horn but managed to limp into Port Stanley in the Falkland Islands where she was discharged, surveyed and condemned.

However, she remained in that port as a coal hulk until 1936, when, at long last, she was considered to be beyond repair and was towed to a long-earned retirement at Sparrow Cove, where she was beached and where she remained until 1970, a multiple record breaker: the first ocean-going iron steamer, the first screw steamer to cross the Atlantic, the biggest ship of her time, the oldest steamer in the world: now happily returned to her native Bristol.

To conclude this chapter I would like to quote some extracts from the Chief Engineer's log of the P. & O. Steamship Company's steamer *Himalaya* whilst on service as a transport during the Russian War in 1854. She was built in 1853 as an iron three-masted sailing-steam passenger ship of 3438 tons. Originally designed as a paddle steamer (remember the Admiralty's predilection for paddle-wheels to carry the mails), she was changed to screw drive whilst building. Her engines were supplied by John Penn & Sons, of the horizontal-trunk type, with two cylinders 84 inches bore by 3 feet 6 inches stroke driving a two-bladed single-screw propeller 18 feet diameter by 28 feet pitch, weighing 7 tons. This engine developed 2000 HP at 56 rpm giving 14 knots on trials. Steam at 14 lb was supplied by four sheet-flue boilers of box shape.

At the outbreak of the war the *Himalaya* was chartered by the Government for service in the Mediterranean, and Mr Gray, her Chief Engineer, kept a complete and most conscientious log-book of the day-to-day happenings in his engine room.

These events are interesting to an engineer on their own account, but they are especially revealing in highlighting the deficiencies of the marine engine in the 1850's, even in large passenger vessel where, presumably, no expense had been spared in the building and fitting out. They are also salutary to the modern marine engineer, keeping watch in his air-conditioned office on the heart of an automated engine room, in showing him that what would nowadays break his heart his great-grandfather accepted as a matter of routine.

The log starts in July 1855 and carried on until September 1856, and the extracts are included by courtesy of the Institute of Marine Engineers:

26.7.1855—Lit fires at 4 am—Steam up at 6.30 am. Started out of Southampton Basin at 8 am—Came to anchor—Touched up cranks, etc. Found things nearly right.—Started 4.40 pm—Full speed at 5.10 pm—Stopped twice to tighten up cranks—rather ill to keep from heating—the cranks keeping better.
28.7.1855—Leaving Liverpool—Stop'd to discharge pilot at 10.30 pm. Started 10.55 pm—at 11.30 pm the after crank warm—Stop'd and slacked it back.
1.8.1855—Obliged to lock off bilge suction pipes as the pumps would not work after about 36 to 40 buckets per watch.
3.8.1855—Stop'd to tighten cranks and trunk gudgeons a little.
5.8.1855—Gibraltar—overhauled fresh water condenser (distiller)—Steam up at 7.10 pm. After starting one of the first coupling bolts broke—stopped and took the pieces out and went away with the five remaining.
9.8.1855—Malta—Got on board from Dockyard a large case with a spare thrust bearing bush sent from Messrs Penn & Company some months ago but never received till now, fitted with lignum vitae according to Mr Penn's instructions.
13.8.1855—All hands up scraping flues—Laid and lit fires at 11 am. Stop'd and slacked the fore engine crank a little.
23.8.1855—Sebastopol—Took a template on shore to get a pin made for the first coupling broken at Gibraltar.
1.9.1855—Turned engines getting trunks touched up.
8.9.1855—Kerch—Men pumping out bilges and turning round the engines by hand.
9.9.1855—Five men turning round engines and scouring the trunks.
12.9.1855—The after port boiler commenced leaking very badly at the back end.
21.9.1855—Off Gozo—temperatures: engine room 125°, stokehold 132°.
1.10.1855—At Genoa—packed trunk glands, examined trunk bearings.

15.10.1855—Engines rather ill to keep from heating—stop'd, cooled main shaft, tightened after trunk journals.
21.11.1855—Men in boiler—On examining air pumps found fore engine piston or plunger all broke round the centre or boss for rod—took it out and put in a spare one that was on board (weight of spare one 322 lb, the weight of broken one 266 lb, difference of weight 56 lb).
15.12.1855—At Constantinople—Examined expansion joint main steam pipe as there was no keeping of it tight—found when packing was out that the cause was that the spick end of the pipe was not cut to the bevel of the casing, 1¼ inch short at one side exclusive of allowance for expansion and not more than ⅞ inch within the faucet.
28.12.55—At Malta—Got off the large steam pipe for repair on shore.
4.1.1856—The thrust bearing wearing away very fast. Stopped at 6.30 put back the thrust bearing a little.
22.1.1856—At Genoa—Lit fire in star'd aft boiler to dry ship's engine room etc., all damp with rain.
26.1.1856—Lit fires at 5 am, steam at 6.30—two mud hole doors gave way and leaked very badly—obliged to blow water out of them and rejoint them. Started for good at 7.50 am, stopped at 10 am to tighten the crank and trunk bearings.
31.1.1856—The steam pipe got full of water and overcharged the safety valves—blew the mercury out of after boiler steam gauge—when drawing fires which caused me lift the safety valves which discharged all the water out with force but wet many of the men on deck and stoke hold.
1.2.1856—Made a new iron strop to try and stop a steam leak on main steam pipe top of the fore cylinder.
5.2.1856—Having no mercury on charge took the mercury out of gauge on top of cylinders to put in on the boiler as being of more use there.
8.2.1856—The fore engine crank got warm, obliged to stop and slack it back at noon—at 4 pm crank cool.
14.2.1856—The large outside ring or slip of lignum vitae broke off at outer end of bush—stopp'd at 7 am. to tighten up crank and trunk bearings, both engines stopped and set back the thrust block as the shaft was working forward and the engines rather ill to keep. . . . . .(?).
18.2.1856—Ship rolling much—throwing much water out of boilers on trunks and pistons.
19.2.1856—Engines racing, obliged to throttle at times—stopp'd to tighten fore engine crank, etc.

5.5.1856—Stopp'd at 5.30 to fix oil cup—touched up both cranks a little—the after trunk bearing mid frame and put back thrust block $\frac{1}{8}$ inch—started 6.20 am—the large wood ring of thrust breaking out.
6.5.1856—Stop'd at 7 am and replaced the lignum vitae slip in the thrust block—Started at 8 am. All going well except the wood bearing wearing away very fast.
7.5.1856—The thrust wearing so fast, obliged to ease the pressure so as to make the wood last into the port of Malta.
8.5.1856—Took off the top of thrust block and lined the lower half of the bush to keep it from working in the block—fitted a new set of lignum vitae slips in with brass rings alternatively to try the effect and see how they might wear.
15.5.1856—Leaving Malta for Gibraltar—started out of harbour at 5.15 am—stop'd at 5.50 am tightened up after trunk journal—started 6 am—noon, stopped and tightened after crank bearing a little—the brass and wood wearing well.
20.5.1856—At Gibraltar changed the brass and wood in the thrust block bearing and put in a new set, keeping the set put in at Malta, which was only about $\frac{1}{10}$ inch, as spare to be placed in at sea should the new wear too much before the end of the run.
24.5.1856—After crank keeping warm, obliged to ease the engine a little.
25.5.1856—After crank ill to keep from heating—stopped and tightened up.
29.5.1856—Stopped engines to tighten up cranks, etc.
6.7.1856—The crank, after engine, still ill to keep from heating—stop'd and tightened up trunk and crank bearings. At 3.30, Hood, a fine man, departed this life in a fit brought on by heat and drinking cold water.
7.7.1856—Stopped to tighten the bushes at 8 pm—eased them as slow as we could get them to work.
13.8.1856—Ship's engineers at after piston fixing air-pump rod in same—men scraping bilges—Factory men taking out the thrust shaft off propeller etc.

Accompanying the log extracts was a plaintive little letter from Mr Gray to his employers dated September 1856 requesting leave of absence to visit his people in Scotland and pointing out that he had received no leave at all since April 1851, and had never had long enough leave to visit his family since joining the Company in 1837.

So much for the good old days: the modern marine engineer in his air-conditioned control room may not be in such close *rapport* with his engines as did his forefathers but he certainly has the best of the bargain.

6 Compound expansion

It was stated in an earlier chapter that our modern steam reciprocating marine engines stem in a direct line from Newcomen's and Watt's early pumping engines, and while this is substantially true, it is not the whole truth. The connecting link between the present-day quadruple-expansion engine, the early paddle engine of the 1820's and the earlier atmospheric engine was the steam which in one case was at 250 lb and 460°F, in the other at 4 lb and 225°F, while the atmospheric engine worked at a vacuum of 20 inches of mercury and a temperature of 160°F. In each case the working medium was the same; the modernisation of the steam plant over these two hundred years lay in the gradual increase in steam pressures with consequential increase in temperatures, leading to lighter, faster engines, increased efficiency and very much reduced fuel consumption.

Steam is rather a peculiar gas inasmuch as it can be used to drive an engine in three entirely different ways: as a means of creating a vacuum in a cylinder, as a pressure in the same cylinder or as a high-speed jet in a turbine. In each case the principle of working is different, and in most engines two, sometimes all three, of these properties are utilised in the one system.

The pioneers of steam, however, knew of one method only by which this fluid could be made to perform work, and that was to introduce it into a cylinder, allow it to expel all the air through the snifting valve, then let it condense into a few drops of water and so create a very good vacuum—the air acting on top of the piston did the rest. Hence the steam performed no useful work on the piston, all it did was to remove the atmospheric pressure from below it, in short, a form of scavenging. For this purpose real steam as we engineers know it was not necessary, plain flabby water vapour was just as effective and initially, the boiler was merely a large copper with a hand lid on the top to check the water-level and fill up as required.

Many years had to pass before a pressure boiler was made, partly because it was not required, but mainly because throughout the eighteenth century no means had yet been devised of building a pressure vessel. Iron plates had to be forged individually from built up scrap, a laborious and expensive business. Again, several plates could be riveted together at their edges to form a vessel but this would barely be watertight, certainly not pressure-tight: the art of boiler making had not yet arrived.

In due course engineers found that by building boilers capable of withstanding a pressure above that of the atmosphere, they could use the steam to push as well as pull, i.e. act in a positive manner, and the steam engine proper had arrived, towards the end of the eighteenth century. The whole of the nineteenth century was devoted to improving the pressure and performance of steam boilers, and steam at 25–30 lb was still being used in the 1860's. The *Great Eastern*, the biggest ship of the century (1858–89), used steam at only 25 lb in her gigantic engines. It was not simply a question of making the boilers stronger or heavier, they were already fairly strong and far too heavy even when built for 10 lb. It was as much a problem of design and compromise. The ideal shape for a pressure vessel is a hollow sphere or a cylinder with dished ends. However, the engineers of the period had to get their boilers fitted into a small boat with the minimum of wasted space, and consequently built them box-shaped with flat sides—really a matter of stowage rather than design.

There were many reasons put forward for the popularity of the low-pressure boiler, one of these being the difficulties in checking steam pressures and water-levels. The steam gauge was a mercury-filled U tube in direct connection with the steam space. A small iron disc carrying a pointer floated on the mercury at the open end and registered the height of the column, equated to the steam pressure, on a graduated scale. If the gauge was shut off and then opened suddenly there was risk of some of the mercury being blown out of the tube with a resulting low-pressure reading on the scale. The water-level was decided by a pair of steam and water cocks spaced a couple of inches above and below nominal water-level. Sometimes it was shown by a float inside the boiler attached to a piano wire through a stuffing box on the boiler top, running over a pulley on the deckhead and down to a scale on a convenient bulkhead. None of these contraptions was dependable and the long, long list of boiler explosions in the first half of the nineteenth century can be traced directly to their fallibility.

It was only in 1847 that Sydney Smith marketed his dial and pointer-type pressure gauge, using a corrugated diaphragm, and in 1850 that Bourdon brought out his flat tube gauge, so that it was many years from that date before mechanical pressure gauges became popular. In the 1850's locomotive boilers were still not fitted with pressure gauges; the safety valves were supposed to do the thinking for the driver. And, of course, although these early pressure gauges were better than the mercury column, they were not entirely trouble-free themselves. At an engineering exhibition in 1869 an invitation was extended to local manufacturers and mill-owners to bring their pressure gauges along and have them calibrated and adjusted free of charge. Most gauges tested were 10% out in their reading, and one, for a boiler working at 50 lb pressure, was 30 lb out on the wrong side.

Perhaps the weightiest arguments on the side of low steam pressures were public opinion and official decree. In 1817 a Parliamentary Commission decided that boiler steam should not exceed a maximum of between 2½ lb and 6 lb pressure, according to type of boiler. Furthermore, that there would be no saving in the use of high-pressure steam, and that boilers should be made of cast iron as this material could be made thicker than wrought iron. In 1839 a Parliamentary Enquiry reported that there was no economy in the use of high-pressure steam. In 1872 the consensus of opinion among marine engineers and shipbuilders was against compound expansion in steam, and McFarlane Gray, of the Board of Trade, had expressed serious doubts on steam pressures over 70 lb in merchant vessels.

Parliamentary Commissions and public opinion were right for once; they were, after all, merely voicing the bitter experience of marine engineers honest enough to acknowledge the limitations of marine machinery during the first half of the nineteenth century. The marine boiler, working at over 25 lb pressure, was inherently unsafe and would remain so until a substitute was found for the jet condenser, where the exhaust steam was condensed by mixing with sea water, the result being pumped back again as salt feed into the boiler. Although Hall had invented his surface condenser in 1839, it took another generation of engineers to appreciate its value and how to use it: the common practice of injecting large quantities of tallow into the cylinder and valve chest may have been beneficial to the working parts of the engine but the eventual clogging of the condenser tubes soon rendered a surface condenser unusable.

Now the salt in sea water, of normal density 5 ounces per gallon, will have little effect on the boiler surfaces until the boiler content, by prolonged evaporation, has reached a density of 34–35 oz when the boiler is fully salt-saturated and the salt deposits on the heating surfaces as a mass of brilliant white crystals. But long before this state was reached our Victorian engineers used to blow down the boilers and freshen up with sea water. In the case of very low-pressure boilers, where the steam pressure was less than the head of water outside the ship, a brine pump driven from the main engine pumped continuously from the boiler, whilst the feed pump topped it up, thus keeping the water reasonably low in density. This operation involved a serious and inevitable loss of heat, but, more important still, it increased the amount of solids in solution brought into the boiler and hence the risk of scale formation. By blowing down and pumping up, the density was kept down only at the cost of increased incrustation—there was no easy way out.

Scale formation was the crippling complaint of the early boilers, and, in the case of high-pressure boilers, one for which there was no cure as long as salt feed was retained. Apart from common salt and magnesium sulphate, both of which remain soluble under normal steaming conditions, sea water contains two ingredients inimical to boiler surfaces, both of which form incrustations on the heating surfaces and have to be removed with great expenditure of time and labour. Carbonate of lime deposits at below boiling point of water and thus would be found in any boiler working at any pressure and, incidentally, with either fresh or salt feed. But this scale is of a chalky nature and can be brushed or, at a pinch, scraped off. Sulphate of lime, on the other hand, forms a hard, rock-like scale that can only be chipped off, and once a tubular boiler was truly scaled up, it was easier and cheaper to renew the tubes than to endeavour to chip them clear. Fortunately for marine engineers, sulphate of lime scale deposits only at about 25 lb pressure, below that pressure it remains soluble in the water.

Incrustations in a boiler are extremely important as they prevent heat transfer from the furnace, flues or tubes to the water and increase the fuel consumption. Thus a scale thickness of $\frac{1}{16}$ inch on the heating surfaces will increase the consumption by 10–15%, $\frac{1}{4}$ inch by 50% and $\frac{1}{2}$ inch by as much as 150%, leading to overheating, collapse and possible explosion.

From the above it will be seen that 25 lb was about the limit of steam pressures until the surface condenser became accepted.

However, not even Parliamentary Commissions can stop progress, and steam pressures had to be stepped up to increase the power-weight ratio and reduce consumption. One of the earliest experimenters in this field was Jonathan Hornblower, who, as early as 1781, increased the output of the old atmospheric beam pumps by adding a HP cylinder alongside the original cylinder, the steam flow being in series from one to the other. Simpson used a similar arrangement of compound cylinders in his rotary beam pumping engine, in each case the HP piston being attached to the beam nearer the fulcrum and therefore having a shorter stroke. MacNaught, another enthusiast of high-pressure steam, made quite a name for himself in the early part of the nineteenth century through increasing the power output of all sorts of steam engines by adding a high-pressure cylinder in series with the existing low-pressure system and, of course, replacing the old boiler by one of suitable scantlings for the contemplated pressure.

These experimenters, however, thought in terms of steam pressure that could be obtained by the existing types of boilers, and the term high pressure was admittedly relative. But people like Oliver Evans, Arthur Woolf and the Perkins family thought of steam not as a flaccid, unsubstantial vapour but as a fiery, elemental force which, once released from the boiler, could provide unlimited power. In this they have since been proved right but not in their day; the materials were wanting. Jacob Perkins was experimenting with high-pressure steam as early as 1823 and his son Angier March, and later his grandson Loftus, thought up boilers capable of steaming at 500 lb. These boilers, however, were not successful in that they lacked proper circulation and their engines failed because steam at such a high

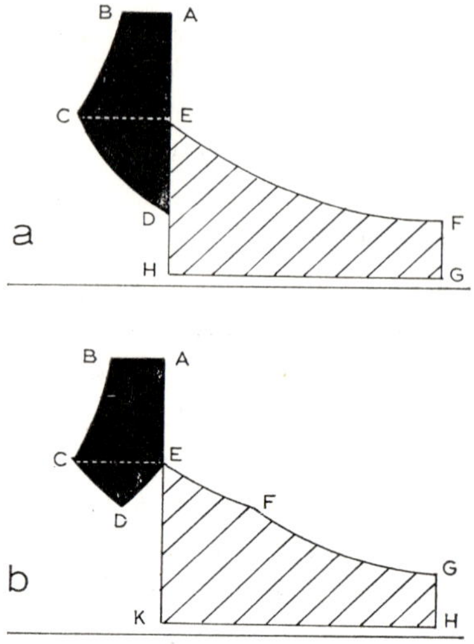

28 Theoretical indicator diagram for:
(a) Woolf compound engine
(b) Receiver compound engine

pressure is not a suitable medium for a reciprocating engine. The steam turbine which could have used such a pressure had still to be invented.

An outstanding name among the advocates of high-pressure steam was Arthur Woolf who introduced the expansion of steam in two stages: compound expansion. In his original design, steam at boiler pressure was partially expanded in the HP cylinder, then exhausted direct into the LP cylinder, the steam continuing to expand in both cylinders on the return stroke. The steam in the HP cylinder was expanded from boiler pressure to condenser release pressure and the cylinder was thus subject to the full range of temperature fluctuation from boiler to condenser at each stroke. In the theoretical indicator diagram for the Woolf engine area, ABCDA represents the work done in the HP cylinder, and EFGHE that performed in the LP. It will be noted that the release pressure C in the HP cylinder is the same as the initial pressure E in the LP cylinder, and that the terminal pressure in the HP is the same as the release pressure in the LP.

This type of engine required the HP and LP crank angles to be either at 180° or 0°, i.e. tandem or annular, and was dependent upon one common slide valve for the two cylinders. A modified type of Woolf engine, the tandem engine with the HP cylinder separately mounted on top of the LP and with separate slide or piston valves to each, became very popular in the late nineteenth century, especially in the North Atlantic passenger trade, where several tandem units could be coupled together on the one crankshaft to form a truly massive and powerful engine.

Another type of compound engine had an intermediate receiver between the valve chests of the two cylinders. In this engine the HP steam was exhausted into the receiver and taken as required by the LP cylinder. The correct expansion did not depend on phasing of the two pistons: the cranks could be placed at 180° or 90° at will. This meant easier starting of the engine and smoother running. The compound receiver engine in due course became the modern triple-expansion engine with cranks at 120°, the workhorse of the Merchant Navy for fifty years and the best known of all marine steam engines.

The indicator diagram for the receiver engine is slightly different from that of the pure Woolf cycle because of the different setting of the valve gears, and here it might be mentioned that the receiver in most compound engines is simply the steam pipe connecting the two cylinders. In the figure, ABCDEA represents the work done in the HP cylinder and EFGHKE that in the LP. In the latter cylinder E is the LP steam at receiver pressure, F the cut off. In the HP cylinder, release to receiver pressure occurs at C and with the LP valve open at the same time, the steam continues to expand from C to D in the HP cylinder and from E to F in the LP cylinder. At cut-off in the LP engine, the HP exhaust is compressed to the receiver pressure E.

A third type of compound engine was the three-cylinder combination of one HP and two LP cylinders, all more or less of the same diameter, really an artifice to cut down the size of the LP cylinder, and one quite often seen in the larger engines, where the diameter of the LP cylinder would otherwise result in a very heavy and unbalanced engine and difficult castings.

The principle of compounding the expansion of steam through an engine is really a logical engineering process and can be compared to walking up a flight of stairs to the first floor instead of climbing up the roan pipe—the effect of lifting the weight of the human body from ground level to first-floor level is the same in each case but the stresses are different. In theory, it is the LP cylinder dimensions that determine the power output of an engine. With steam conditions and revolutions equal, a single-cylinder engine with given bore and stroke would supply the same horsepower as a compound engine with the same LP bore and stroke.

Theory, however, is wantonly prodigal of the ship-owner's hard-earned money and can afford to neglect excessive condensation, weight and unbalance. In a single-cylinder engine using high-pressure steam a very early cut-off would be necessary to get the required number of expansions, and in order to use the steam economically, with a release at as low a pressure as possible, a very large cylinder would be required to accommodate such a volume of low-pressure steam. The working parts would be subject to temperatures ranging from that of high-pressure steam to that of the condenser, involving undue thermal stresses in the castings and excessive initial condensation. The mean effective pressure in a cylinder with such a pressure drop from inlet to exhaust would be small compared with the initial high-pressure steam. This would entail a large piston strong enough to withstand high initial pressure and these characteristics would be reflected in all the running gear of the engine: very strong and heavy to stand up to high initial shock.

All the above would combine to make an extremely heavy and costly engine with a heavy steam consumption due to high condensation losses, low power output due to friction losses in the large-sized bearings and rubbing surfaces and considerable cyclic variations in revolutions. To rectify the latter disadvantage, a flywheel commensurate with the engine size would have to be fitted, thus increasing the engine weight still more.

By breaking down the expansion process into two or more stages in series, the cylinder volumes can be reduced to reasonable sizes and the initial stresses on the different pistons will be reduced, while a more uniform turning moment will be obtained. More important still, however, is that the initial condensation will be considerably reduced, and only in the LP cylinder is the condenser temperature evident. With the steam expanding in succession, through two, three or more cylinders in series, the temperature and pressure range is fairly small and there is a certain amount of re-evaporation of condensation during the intermediate exhaust as the steam flows through the engine.

One of the earliest examples of a steam engine working on the pure Woolf cycle was built at Leeghwater in the 1840's to drive a number of pumps for draining the Haarlem Lake in Holland to create a polder—a tract of arable land below sea-level. This pumping engine was of the vertical annular type, with cylinder diameters of 84 inches and 144 inches, one inside the other, and a stroke of 10 feet. The central piston rod of the inner piston and the two piston rods of the annular piston were coupled to an overhead crosshead plate, with a central guide rod. Eleven beam pumps in a circle surrounded this steam engine and the ends of their beams, fitted with rollers, engaged below this crosshead plate and were worked up and down when the engine was in motion, as in a simple old-fashioned beam pump. The action of the pump was as follows: steam was admitted below the central piston moving the crosshead plate upwards and allowing the pump buckets to descend into their respective wells. At the top of the stroke an equilibrium valve opened and placed both central HP cylinder and annular LP cylinder in communication, while the underside of both pistons was in communication with the condenser. The exhaust steam acting on the top of both HP and LP pistons, now working as one unit, pushed the crosshead plate downwards, thus raising the eleven buckets on a 10-foot stroke.

The steam in this pump engine expanded six to eight times its original volume, and to cope with the large cylinder volumes, a pair of air pumps were worked from the beams, each 40 inches diameter by 5 feet stroke.

Compound expansion, like everything else associated with the steam engine, started on land and became established on river craft long before it was accepted for sea-going ships. Although foreign trade ships started using compound engines in the late 1850's it was not until the 1870's that they became popular, while cases are on record of compound engines being used to drive river steamboats built around 1820–30, although in some cases the engines were compounded by adding a high-pressure cylinder or by renewing the engine entirely.

One of the earliest recorded steamboats with a compound engine was the *James Watt*, a small side-paddle steamer originally fitted with two steam cylinders 20 inches bore by 40 inches stroke. In 1829 she was reconstructed and equipped with a new engine, her name being changed to *Stad Keulen*. This new engine consisted of a pair of compound units each consisting of a diagonal HP cylinder built up on a lattice of crib work in the engine room proper, and a horizontal LP cylinder placed on the deck, the drive being of the Gorgon type: piston rod, crosshead, connecting rod, and both HP and LP pistons worked on to the same crankpin. This engine, apart from being compound, had some unusual features about it not met with in later engines. The crosshead guides ran in rollers on both HP and LP engines, and the LP piston drove an air pump in tandem on an extension of the piston rod. The slide valves were worked, not by the usual eccentric but by a cam plate on the shaft, and the motion was communicated to the valves by means of a square frame round the cam and a lattice eccentric rod, an arrangement reminiscent of steamboat practice in the southern states of the USA for that period.

To complete the unusual note, the wooden paddle floats were not made flat but were stepped upwards as in a flight of stairs.

Another unusual conversion to compound drive was the steam tender *Hercules*, built in 1829 for the Nederlandsche Stoomboot Maatschappij at Rotterdam, for service on the Rhine and Maas. The original engine consisted of a pair of horizontal units placed on the deck driving a lay shaft geared up to the paddle shaft, an odd blend of the old-fashioned geared paddle drive with the up-to-date compound engine, but then conversions nearly always show some form of anachronism.

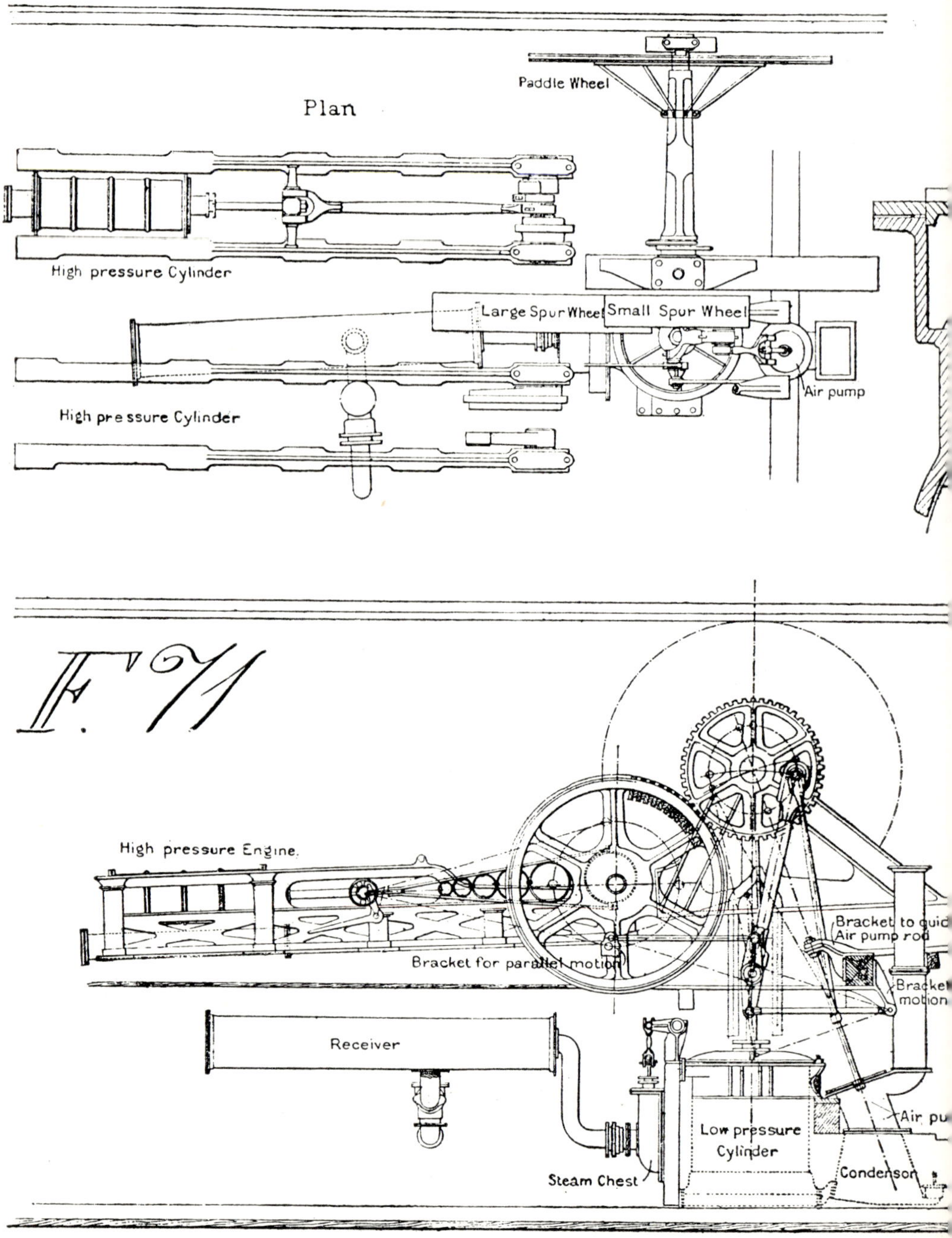

29 SS *Hercules* (1829)—Arrangement of compound engine and boiler

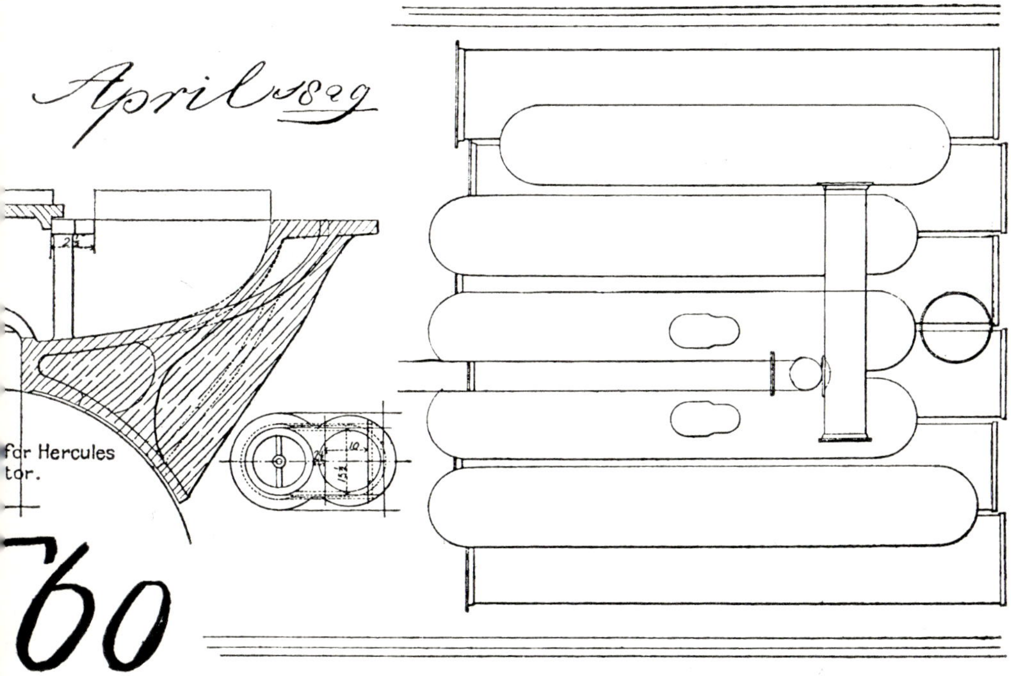

Hercules.

Side elevation of engines and longitudinal section of boiler

Scale

12 9 6 3 1 2 3 4 5 6 7 8 9 10 11 12 13 14 15 16 17 Eng feet

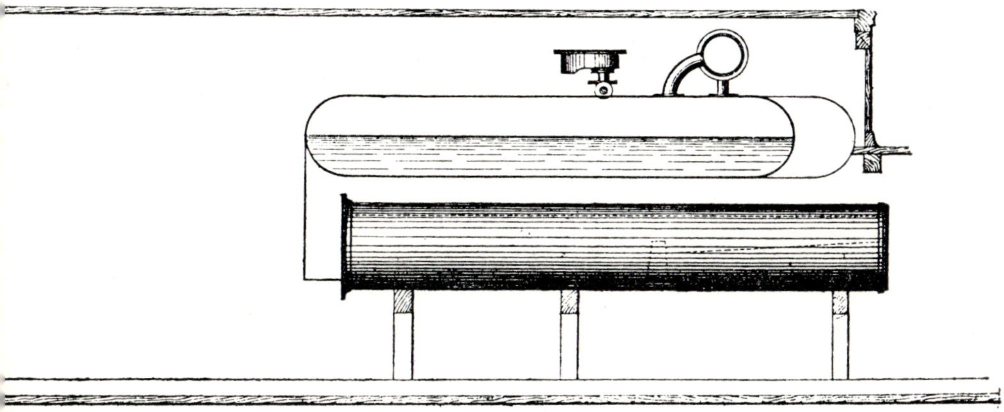

The two steam cylinders were retained, and a large-bore cylinder removed from some side-lever engine was placed vertically in the engine room below the paddle shaft and coupled to a crank thereon. The gear ratio of the engine shaft to the paddle shaft was roughly 3:2, consequently the two HP pistons made two strokes to every three of the LP. A new high-pressure elephant boiler was fitted at this time, giving a pressure of 70–80 lb, an exceedingly high pressure for those days, and this was substantially renewed in 1841 by a larger more modern boiler still.

One of the best known of the early compound engined steamboats was the water-bus *Cricket* plying between Hungerford Bridge and London Bridge, and owned by the Halfpenny Steamboat Company. She was built in 1846 and fitted out by Joyce and Co. of the Greenwhich Ironworks with a pair of Woolf compound oscillating engines driving a straight-forward paddle shaft. The HP and LP cylinders of each engine were bolted together to form a common unit oscillating as one, and their piston rods laid on to the same crankpin, made exceptionally long to span the two-cylinder centres. The crankpins on the port and starboard engines were set at 90° to each other, and the air pump was placed between the two in the usual way. The valve gear, of the single eccentric type, was that normally fitted to oscillating engines, but in this case it had to drive two slide valves on the same stroke from the rocking shaft and thus had to be so much more robust.

This engine was supplied with steam at 20 lb by two tubular boilers placed side by side, each with a single furnace and about seventy return smoke tubes. In late 1847, leaving Adelphi Bridge with 150 passengers on board, a main longitudinal stay in the starboard boiler endplate failed and the endplate blew out. The force of the escaping steam shot the boiler through the after part of the vessel and the main cabin, killing seventeen persons outright and injuring sixty. At the enquiry the stated cause of the explosion was that responsible for most of the boiler mishaps in those days: overloading of the safety valves.

This disaster ended the career of a most economical and speedy little boat; she could make 14 knots at 4½ lb/IHP/hour, considered a good coal consumption for that period.

The protagonists for the compound principle on ocean-going ships were John Elder and Alfred Holt. The *Brandon* was built in 1854 by Randolph Elder & Co., on the pure Woolf cycle, and she was followed by the Pacific Steam Navigation Co's *Valparaiso* and *Inca* in 1856, all three working at 30 lb steam pressure. The P. & O. Company brought out their *Carnatic* in 1865 driven by steam engines with cylinders 43 and 96 inches bore by 3 feet stroke, working at 26 lb boiler pressure. These vessels were most economical by the standards of the period,

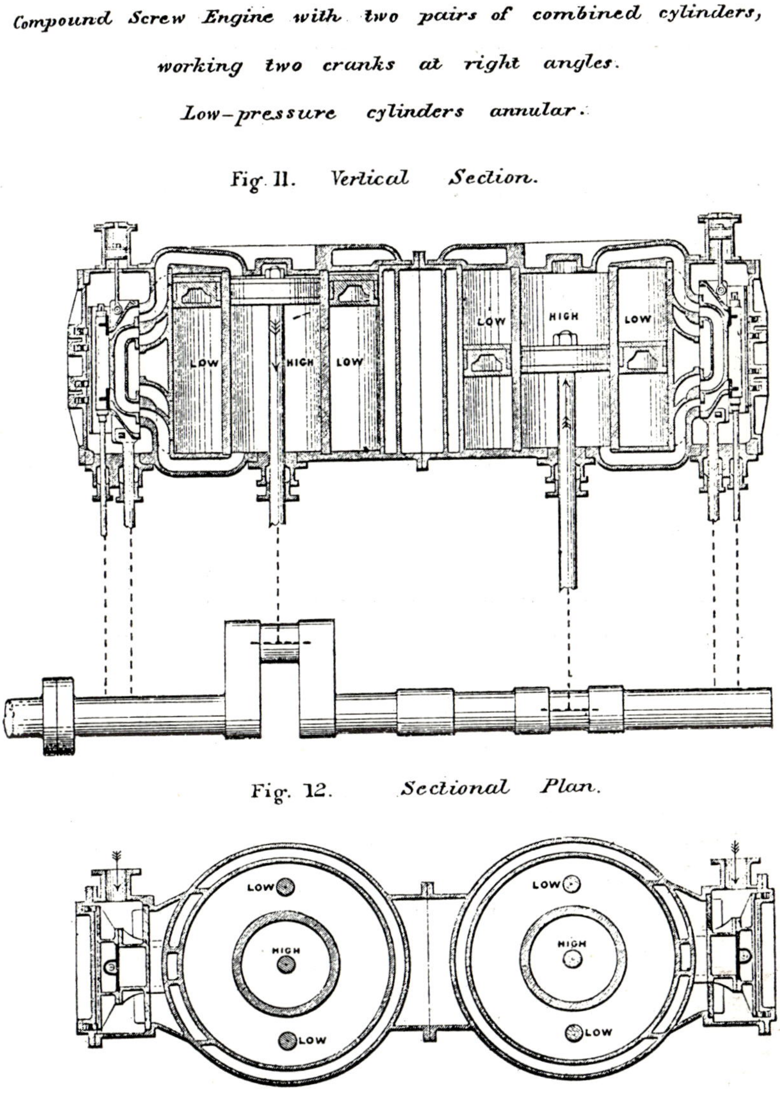

30 Annular engine with two cranks

giving a steady consumption of some 2 lb/IHP/hour. These early compound engines were mainly of the tandem type, with the HP and LP cylinders placed one above the other, both pistons working on the same rod on to the one crankpin; in some cases the LP was on top, in most designs, however, the HP was placed upper-

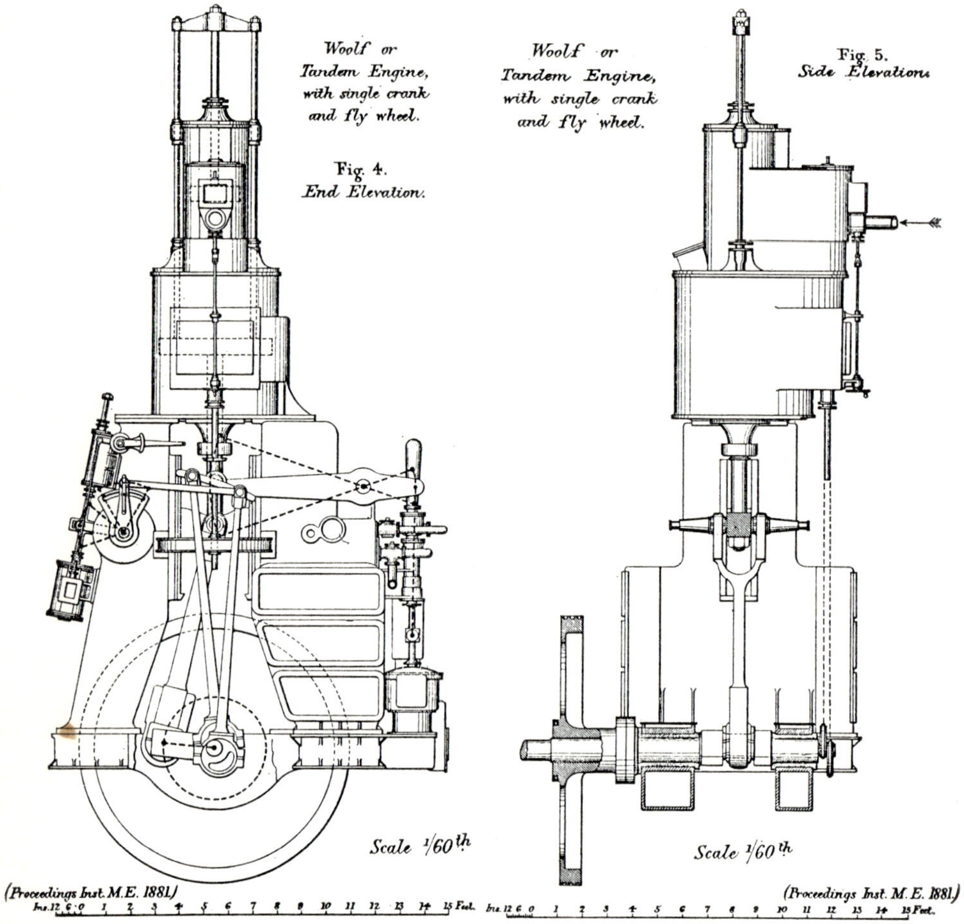

31 Alfred Holt's single-crank tandem compound engine (1862)—End and side elevations

most. These tandem engines came to be very popular in the latter part of the nineteenth century, and in the speech of the day the HP cylinder was invariably referred to as the monkey cylinder. Another type met with in the earlier compound engines was the annular piston type, with the HP in the middle and the annular LP surrounding it, their piston rods either being coupled together to the one cross-head (i.e. with cranks at 0°), or else the HP driving one crank and the LP driving another at 180°; this phasing was important in the pure Woolf cycle.

It may be recalled that the *Flora*, built in 1862 and one of the first twin-screw steamers, had a pair of annular engines with the HP and LP piston rods coupled through a common cross-head to the one crankpin. Another early steamer, the *Jose*, originally fitted with a pair of single-expansion cylinders 46 inches in diameter with cranks at 90° and working at 15 lb boiler pressure, was converted to compound expansion by substituting an annular engine working at 48 lb for the original unit. The new engine had cylinders 22 and 52 inches diameter with a 3-foot stroke, the 52-inch LP annular piston having an area equivalent to a normal piston 44½ inches diameter. At the same time, the jet condenser was discarded in favour of a surface condenser, and incidentally the vessel was lengthened some 30 feet during the conversion. Tests taken before and after conversion showed a net gain in fuel consumption from 4·6 lb to 2·12 lb/IHP/hour, a pretty conclusive proof of the advantages of high-pressure steam.

Alfred Holt brought out a series of ships between 1862 and 1866 driven by his famous single-crank tandem engine, with boiler steam at 60 lb pressure and a fuel consumption of 2¼ lb/IHP/hour. These were vessels of 320 feet length between perpendiculars, beam 36 feet 4 inches and carrying 3000 tons on a draught of 23 feet. The engines had cylinders 27 inches and 58 inches diameter, with a 5-foot stroke, the HP cylinder being placed above and connected to the LP cylinder. The HP piston rod, actually a tail rod, had an overhead crosshead beam, from the ends of which a pair of long side rods came down outside the HP cylinder, through stuffing boxes in the LP cylinder-top cover and connected to the LP piston. This design did away with the inconvenience of a pair of glands to pack between the two cylinders and reduced the height of the engine by that amount.

The LP piston rod transmitted the full power of both pistons on to an overhung mill-engine-type crankpin. The LP slide valve and the HP piston valve were worked from the same valve gear: a pair of eccentrics with Stephenson's link motion, with a steam reversing cylinder. This engine had a large, single bearing which also acted as a thrust bearing, placed between the shrunk-on crankweb and the eccentric sheaves forged solid with the shaft. Abaft the eccentric gear a 5-ton flywheel some 12 feet in diameter acted as governor to the engine when running and as turning wheel when stopped. It also had another function, that of starting wheel when the crank stuck on dead centres. The flywheel rim had teeth cast in its periphery, and a steam heaving cylinder with an extended piston rod was placed horizontally over the flywheel. This rod carried ahead and astern pawls which could be engaged in the flywheel teeth at will and pull the engine off centres as necessary with just a touch of steam in the cylinders. To prevent possible damage, the starting cylinder was in tandem with an oil cataract or dashpot cylinder, not unlike the reversing gear.

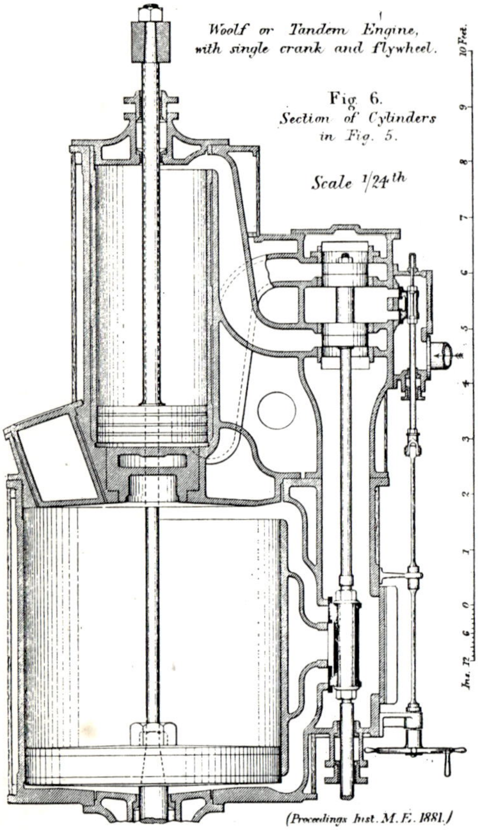

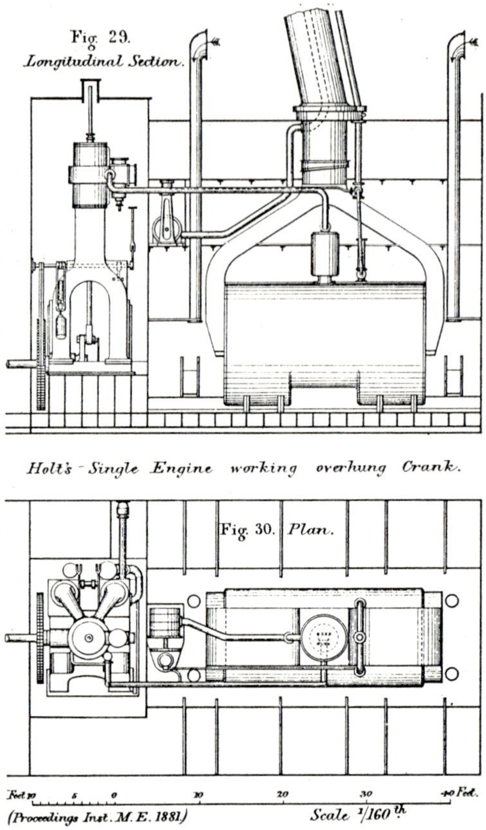

32 *Left:* Single-crank tandem compound-engine valve gear (1862)
33 *Right:* Alfred Holt's single-cylinder marine engine—arrangement of engine room (1881)

This engine drove its own air, feed and bilge pumps direct from the crosshead by link gear, also a small single-acting fresh-water pump from the air-pump rod. In addition to these pumps slung around the condenser, to the steam reversing gear on the port column and the steam heaving gear above the flywheel, a general service banjo pump was fitted on the forward side of the front column. The only independent pump in the engine room, not attached to the engine, was a 6-inch Gwynne centrifugal pump to circulate the three-pass surface condenser, a rare luxury for those days.

Steam to this engine was supplied by one double-ended boiler 16 feet in diameter by 24 feet long working at 80 lb pressure, with a total weight under

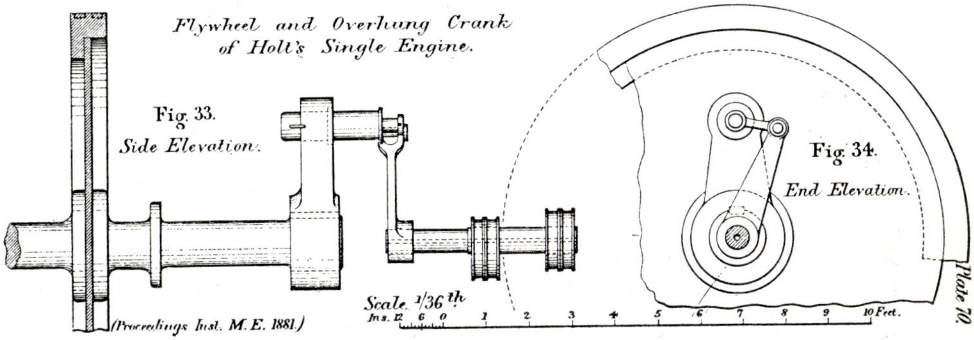

34 Alfred Holt's single-cylinder marine engine—arrangement of crank and eccentric sheaves (1881)

steaming conditions, of 157 tons. This engine developed 1350 IHP with a good coal consumption and gave a ship speed of 12 knots.

Handling such a large, unbalanced single-crank engine would frighten modern marine engineers, and manœuvring the engine for hours on end entering or leaving a large port through a maze of docks would constitute for them the ultimate in nightmares, yet I still remember an uncle of mine, who sailed as an engineer with one of these Holt engines, describing with what simplicity the engine would be run over from ahead to astern, and stating most emphatically that it never stalled on dead centres.

In spite of the success of the tandem compound engine Alfred Holt abandoned the two-cylinder expansion principle while still retaining high-pressure steam, and in 1881 he built a 2200-ton steamer propelled by a high-pressure, single-cylinder steam engine and going as far as to dispense with the crankshaft entirely by carrying the intermediate shaft right into the engine with a single overhung crank shrunk on to the forward end. The eccentric sheaves were forged on to a small layshaft linked on to the crankpin. As will be seen from the longitudinal section of the machinery spaces, the engine was exceedingly lofty and practically filled the very small engine room. The double-ended boiler had a system of induced draught provided by a centrifugal fan in the 'tween deck, delivering a high-pressure air blast through a nozzle in the funnel, on the lines of locomotive practice.

Although Holt had abandoned the tandem compound engine in favour of his single-cylinder high-pressure engine as being even more economical, most ship-owners persisted in the design, as they found it a convenient type for the ever-increasing size and speed of their ships. A tandem compound unit, a complete engine in its own right, could be duplicated or triplicated on the same crankshaft

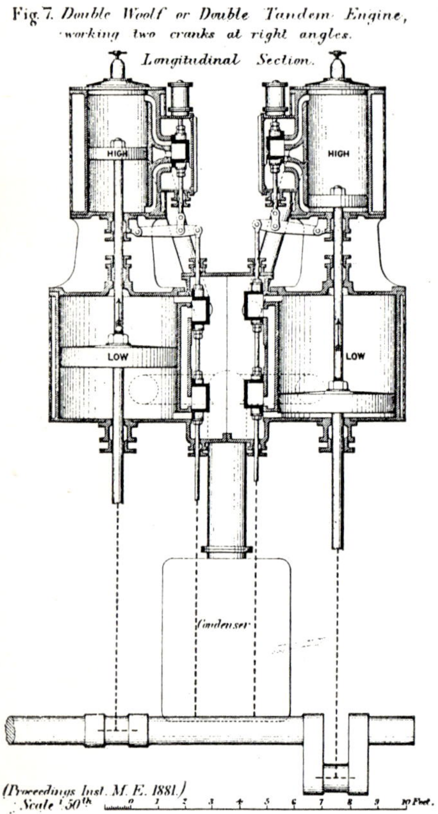

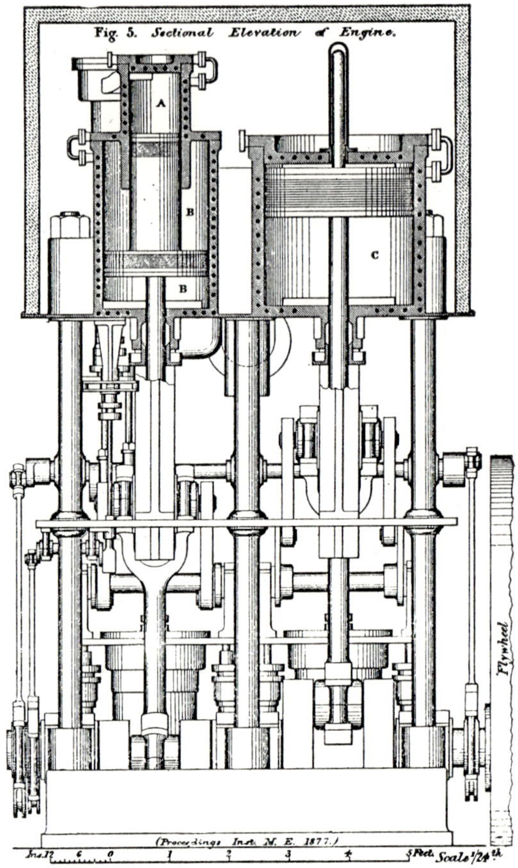

35 *Left:* Section through two-crank tandem compound steam engine—arrangement of valves and valve gear
36 *Right:* SS *Anthracite* (1880). Triple-expansion engine using steam at 350–400 lb. Note steam jacket formed by live-steam coils cast in cylinder walls and covers

to form a very powerful engine indeed. This trend led to the truly massive engines seen in the Atlantic passenger steamers of the 1870's and 1880's, two- and three-crank machines with their HP cylinders towering up into the skylight casing and their huge bulk almost filling the engine room.

This penchant for tall engines inevitably led to modifications in the design of passenger ships and to additions to their superstructures. The earlier steamships, with their horizontal screw engines, had been flush-decked with very few breaks in the sweep of the deck-level: a narrow cabin amidships, possibly a smoke room, a couple of skylights and the odd companionway. Skylights were sited on a shallow coaming, perhaps two feet above the deck, with benches all round. With tandem engines, however, the monkey cylinder protruded above the deck and it was found necessary to extend the engine room into a narrow casing and place the skylight a deck higher. With still greater heights and more bulky cylinders which now all but filled the casing, the latter had to be heightened a further deck height and extended to the ship's sides, eventually becoming a bridge house complete with passenger accommodation.

As pressures increased, it was found convenient to expand the steam in three steps, and Dr Kirk's triple-expansion design came on to the scene in 1880, to be followed a few years later by the quadruple-expansion engine, the latter being generally accepted by the end of the century. There were, of course, variations in the design of the multiple-expansion engines, some of which still retained the lofty tandem principle while others preferred to expand laterally on four cranks, with two LP cylinders. As experience with triple expansion increased, four-stage expansion was introduced and became popular in the 1890's, especially for large passenger vessels.

Just as the triple-expansion engine was sometimes made with four cylinders, so the quadruple engine could be built with five cranks, as in the case of the *Inchmona*, built in 1896 by W. Gray & Co. of West Hartlepool, for Hamilton Fraser & Co. of Liverpool. The main engine had five cylinders 17, 24, 34, 42 and 42 inches in diameter, with a stroke of 42 inches, and worked at a boiler pressure of 225 lb. Such cases were comparatively rare, however, as ship-owners found that the reduction in the LP cylinder size and weight did not warrant the additional cost and maintenance problems, not to mention the loss of cargo space by the added length of the engine room.

During this period of research into the principles of multiple expansion, and of the development and production of triple- and quadruple-expansion engines—what might be termed the normal growth of the steam engine—considerable experiment was being carried out on super-pressure steam, and in 1880 Loftus Perkins brought out his *Anthracite*, a small steamer 60 feet long driven by a triple-expansion engine on two cranks, specially built to run on steam at between 350 lb–400 lb pressure supplied by a tubular boiler of advanced design.

This engine had cylinders 10, 14 and 28 inches bore by 18 inches stroke and developed 125 IHP at 110 rpm with a fuel consumption of 1·72 lb/IHP/hour, the steam pressure varying from 450 lb to 350 lb in the boiler. The HP and IP cylinders were in tandem on the one crank, while the LP cylinder drove the other crank, at right angles to it. The HP and the IP cylinders were single acting, steam acting on top of the HP piston and exhausting to the bottom of the IP cylinder, thus forming a complete Woolf unit. The steam from the IP bottom cylinder exhausted into the LP cylinder via the top part of the IP cylinder, which was in communication with the LP valve and thus formed a receiver.

Poppet valves instead of piston valves were used and a very good vacuum of 28½ inches was maintained by the unusual type condenser. This consisted of vertical closed-topped tubes having open-ended tubes within them. The close-topped tubes terminated at their bottom ends in one tube-plate and the open-ended tubes in another. The circulating water went up through the open-ended tubes and then came down through the annular space between the two tubes, and this arrangement neatly got over all expansion difficulties.

Perkins claimed that the success of his high-pressure economical engine lay in the complete freedom from loss through condensation by a live steam jacket on all cylinders, whereby a reasonable amount of reheat was given to the steam in the lower stages. This jacket consisted of coils of wrought-iron tubing cast into the cylinder walls, bottom and covers through which circulated live steam from the boiler.

This *Anthracite*, small as she was, crossed the Atlantic to North America and returned to Britain and gave a most successful account of herself. Upon her return she was engaged in a number of experiments and trials and it is worth quoting an extract from the report on the *Anthracite*'s trials made by a prominent marine engineer of the day:

> 'Indicator diagrams were taken every half hour until they ceased to put coal on the fire; then they were taken every quarter of an hour; and then, when the engines fell off from their normal speed, they were taken every five minutes until the engines came to a stand. Mr Thorneycroft, who at this time went into the boiler room to see how much fire remained on the bars, found the firebox in complete darkness, and had to take a light to look for what was left of the fuel; this was found to be only a few cinders, so cold that they could be carried in the hand. The engines worked until the effective boiler pressure had fallen to zero, at which time they came to a dead stand, having just before been making from 28 to 30 revolutions per minute.'

One feels that an engine and boiler unit with this performance deserved a better fate.

The popularity of higher pressures with their promise of reduced fuel consumption brought about many conversions in existing ships whose compound

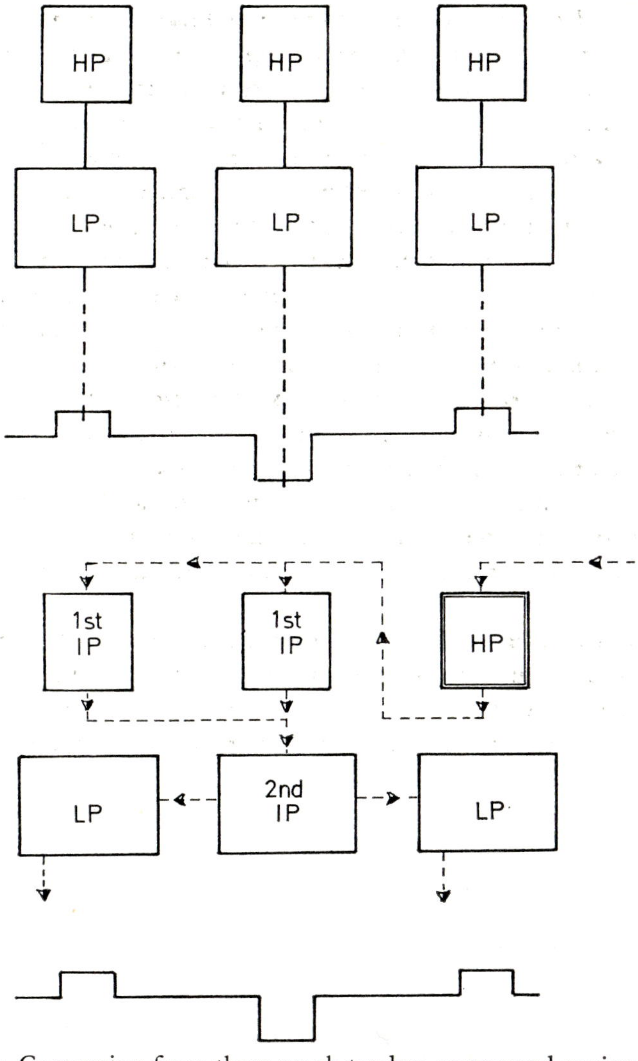

37 *La Gascoigne*. Conversion from three-crank tandem compound engine with steam at 90 lb to quadruple expansion with steam at 165 lb

engines were too good to be scrapped. One of these was *La Gascoigne*, originally driven by a compound tandem engine with three HP and three LP cylinders, on three cranks at 120° to each other, taking steam at 90 lb. She was converted to quadruple expansion by removing the forward HP monkey cylinder and replacing it by a much stronger one of roughly the same bore, built for a boiler pressure of 165 lb. By suitably modifying the receiver pipes between the various cylinders, the steam path through the engine took an entirely different route than that originally intended. From the boilers, the steam passed through the new forward HP cylinder, exhausting into the two remaining HP monkey cylinders (now working jointly as first IP); from these two cylinders it was reunited in the central LP cylinder (now acting as second IP), passing from there into the forward and after LP cylinders and finally exhausting into three condensers each with its own air pump driven by links from the main crossheads.

A more unusual conversion from old to new, or rather ancient to modern, was that of the Tyne passenger ferry steamer *Lady James* which in 1880 was gutted, equipped with a high-pressure propulsion plant and renamed *Loftus Perkins.* The original engine was a single-crank low-pressure grasshopper engine using steam at 35 lb, driving a pair of side paddle-wheels. The cylinder and running gear were removed, retaining the original paddles, shafting and side levers, and a monobloc arrangement of HP, IP and LP cylinders was fitted in their place. The HP and IP cylinders, in tandem, were single-acting, and their pistons were fitted on to the same rod, while the LP cylinder was double-acting. Cylinder diameters were 10, 20 and 28 inches, with a 2-foot stroke, and both piston rods coupled to a common crosshead which in turn worked the side levers.

The old jet condenser had been discarded together with the original cylinders and boilers, and a new surface condenser was installed below the engine and shared the functions of a bedplate.

The boiler was of the Perkin's all-tube type, and had its safety valves adjusted to 450 lb, and on completion the new installation was claimed to be only half the weight of the original. It had been remarked earlier on that a conversion quite often involved an anachronism, and this was very much the case of the *Lady James* where an out-of-date engine was given a face lift with very advanced type steaming conditions, a process comparable to fitting one's lawn mower engine with a supercharger.

The triple-expansion engine of the 1890's remained unchanged in design and construction until after the First World War: the tramp steamer of the 'nineties was almost identical to that of the 1920's. After the war, however, a modest revival of the old Woolf cycle took place, resulting in the Ajax engine in the United States and the Lentz engine in Germany and Scandinavia. The Ajax engine was a tandem or steeple compound engine with the HP and LP cylinders combined in a single unit, the engine being composed of three such units with their cranks at 120°. The common piston rod carried both pistons, and the exhaust valve of the HP cylinder served as the inlet valve to the LP cylinder. The exhaust from the LP cylinder was through ports in the liner uncovered by the piston at bottom of stroke. The Ajax engine was double-acting with reference to the piston rod and crankpin, but was, in fact, single-acting in respect of the pistons. The HP cylinder took steam only on the up stroke and the LP cylinder only on the down stroke. Steam control was by small piston valves operated by bellcrank and push rods from a chain-driven layshaft, and steam conditions were normal cargo-boat practice: 200 lb to 230 lb with mild superheat.

The Lentz engine was originally a triple-expansion engine with double-beat valves worked from a camshaft instead of the more conventional slide valves driven by Stephenson's link motion. In the 1920's this engine was redesigned to become the double-compound marine engine: a pair of two-cylinder compound engines on the same crankshaft. Each compound section had the HP and LP cranks positioned at 180°, on the Woolf principle, and their cylinders were so arranged that steam from the HP exhausted directly into the LP cylinder through a common valve, as in the Ajax engine. The two compound sections were joined together with the HP cranks at 90°.

The characteristic feature of the Lentz engine was its valve gear. In the original triple-expansion engine the steam in each cylinder was controlled by double-beat drop valves worked by cams on individual camshafts, there being one separate camshaft to each cylinder. These shafts were all concentric, the one inside the other, telescope fashion; the LP, being the longest, was a solid shaft with a tubular IP next and the HP outside. Each camshaft was driven by its own loose eccentric sheave on an extension of the crankshaft at the forward end of the engine. The three eccentrics had athwartship racks cast into them which mated with a movable fore-and-aft skew rack through their centres. Movement of the skew rack altered the position of the eccentric sheave, relative to the crank, from ahead to astern. The object of this Lentz valve gear was to change the simple harmonic motion provided by the eccentric into a more precise opening and closing of the valve through a cam profile. The objection to this type of gear, from the maintenance angle, was that the drop valves had to be ground in hot; the chests had to be opened out as soon as steam was shut off the engine.

Cam-operated valve gear was quite often fitted to conventional triple-expansion steam engines on the HP and IP cylinders, sometimes as designed, sometimes as a conversion. In these cases, conventional valve gear was used to work a short camshaft sited between the upper and lower valve chests, the steam valves being either poppet valves or matchbox valves, i.e. slide valves working between the chest face and a saddle.

Finally, just as Alfred Holt in the 1880's abandoned his single-crank tandem compound engine for a single-cylinder marine unit, so in the 1930's the multi-cylinder steam motor with piston exhaust attempted to usurp the established position of the triple-expansion engine. The uniflow engine was a double-acting steam engine where full expansion took place in the one stroke, the steam being admitted through a cam-operated poppet valve and exhausted to the condenser through ports in the cylinder liner uncovered by the piston at the bottom of its stroke. With the power unit applied to a single crank, like a diesel engine, this steam engine could be built up of a number of such units, all identical, to form a large multi-cylinder motor. Its resemblance to the diesel unit was heightened by a closed crankcase, chain-driven camshaft and spring-loaded valves.

This engine was claimed to be as efficient as a conventional triple-expansion engine by reason of its very small clearance volumes and the large ratio of expansion in the cylinder, due in great measure to the absence of condensation on the cylinder walls. This efficiency had to be paid for, however, by the lubrication problems and their effect on boiler feed.

The presence of this uniflow engine among the present-day triple- and quadruple-expansion engines was rather disconcerting to the old-fashioned marine engineers, inasmuch as it upset all their thermo-dynamic theory. In the last hundred years, through experiment, calculation and sheer hard work, they reduced the fuel consumption of the ordinary marine engine from over 4 lb/IHP/hour to well under 1 lb. Starting with a single-cylinder low-pressure steam engine, they convinced the ship-owner (and ship-owners take a lot of convincing) that the future of steam lay in compound engines and high boiler pressures. A subsequent generation of engineers persuaded the ship-owner into triple- and quadruple-expansion machines using still higher pressures.

It must have been rather embarrassing to these engineers to have to admit to their ship-owners that single expansion in the one cylinder was probably just as efficient after all.

7 American river steamer engines

In the preceding chapters it was shown how the marine engine grew from the tiny rickety machines driving river ferries in the 1820's into the massive ponderous monsters that propelled the Atlantic ferries in the 1890's. We have seen how the original engine, the prototype, changed in form and design, passing from paddle to screw, and climbing laboriously up the expansion scale to bring its fuel consumption down from 10 lb coal/HP/hour in 1820 to 1·5 lb /HP/hour at the turn of the century. This gradual refinement followed a definite and logical path, and the marine steam engines built in the industrialised countries throughout Europe and the eastern freeboard of North America all showed similar basic designs and characteristics. In short, at any one period of the nineteenth century, a ship's engine and boilers built in USA would be similar to their counterpart manufactured in Britain, France or Germany, allowing some latitude for design and detail. The British and German engines were usually more down to earth in design, sturdily built but possibly lacking in imagination. French engines showed more elegance and care for style, even when these properties were not required in the design, while American-built engines tended towards gauntness and economy in materials and showed little concern for good appearance.

In its growth and expansion throughout the world, the marine engine had achieved by the end of the nineteenth century a position similar to that of the present-day motor car where a private saloon or a lorry built in Japan could be mistaken for a local Midlands product by ninety-nine Britons out of a hundred. The performance and efficiency being taken for granted, the shape and styling would be in accordance with a universal standard adhered to by all nations.

In North America orthodox engines, identical to those manufactured in Britain and on the Continent, were fitted in foreign-going ships, and from the historical point of view there was nothing particularly noteworthy in their construction. However, side by side with the American fleet of conventional deep-sea ships there existed a vast multitude of fast light-weight steamers of peculiar build plying the thousands of miles of inland waterways and lake shores, steamers such as existed nowhere else in the world, designed to cope with the many problems of big-river transport and thinly populated country, with equipment and engines of an engagingly earthy and home-spun quality.

These unusual steamers fell into two separate categories: the Western River type and the Eastern River type, both distinctive in design the one from the other, and both entirely different from the usual form of river boat.

The Western River steamer was peculiar to the Mississippi–Missouri–Arkansas–Ohio–Red River basin, extending from Pennsylvania in the East and Montana in the West down to the Gulf, draining half the total of present-day USA. In the

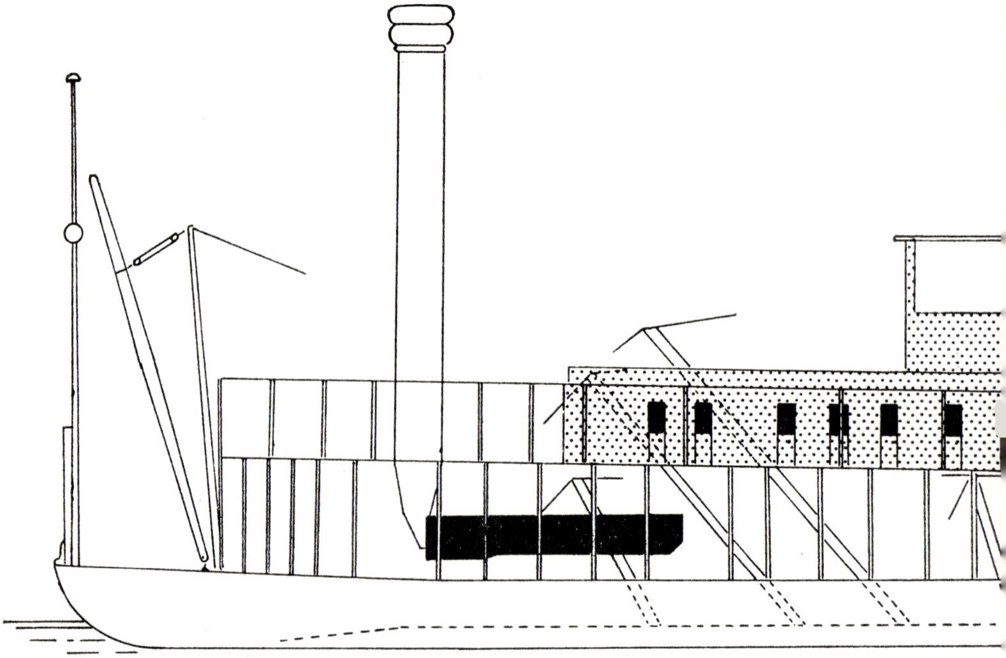

38 Western River steamboat—position of engine and boiler

1850's there were few roads and fewer railways to serve the Middle West, and all passenger and goods traffic went by river. The rivers themselves, both the Mississippi and her tributaries, were usually broad and shallow, with occasional rapids, sandbanks and a temperamental tendency to change course without warning. In addition to these normal navigational hazards were the trees falling into the water through erosion of the banks. These waterlogged trunks would become embedded in the river bottom and present dangerous obstacles to all river craft, the dreaded snags that could rip the bottom out of a ship in a matter of seconds. No special care on the part of the pilot could guarantee immunity from the risk of snagging; if the tree broke the surface he could detect it in time and keep clear of it. If, however, it was well below the surface and caused no ripples to show, his only hope was that his very shallow draught would enable him to sail over it.

This Western River steamer, the Mississippi-type paddler, evolved from the earliest keelboats with the advent of steam. Owing to the ever-increasing river traffic through the opening up of the Middle West, these steamers were built larger and larger, but, because of the general shallowness of the rivers, of their

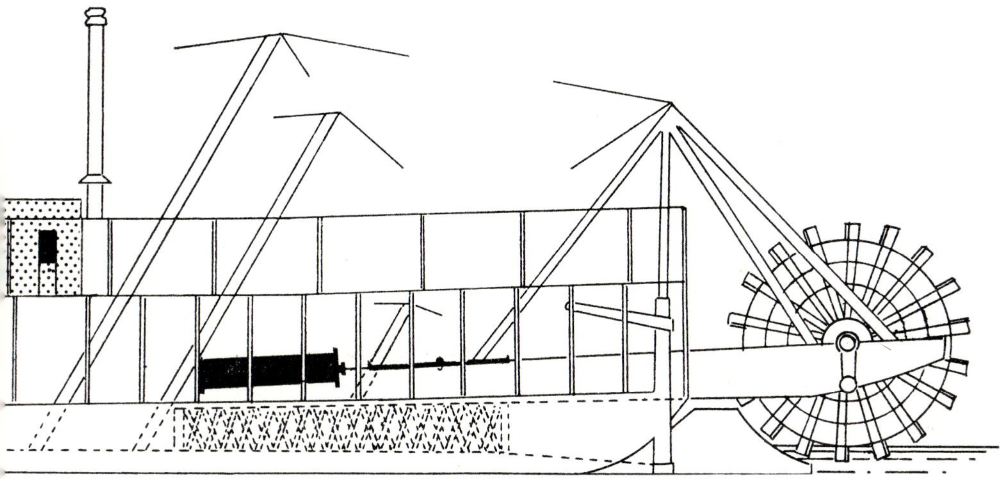

high incidence of rapids, shoals and snags, the hulls could be expanded in two directions only: length and breadth, the draught had to remain unaltered.

The type of steamer that eventually developed was a flat-bottomed ship-shape hull, with little freeboard and less draught where what we could call the weather deck became the open working-deck-cum-engine room, and with two, three or possibly more upper decks supported on tall pillars, containing the accommodation, saloons and officers' quarters. Above all this was the pilot house, placed well aft, containing the huge steering wheel around which the pilot had to climb when navigating a sharp bend in the river. The working deck extended well beyond the moulded breadth of the hull to form the guards, a form of sponson extending for nearly the full length of the hull, thus increasing the effective working deck area by some 75%. The upper decks were made the same width as the lower one, the overhang being supported by ties from overhead beams placed across the deck pillaring. Another peculiarity of these steamers was the method of strengthening the hull by means of hogchains, the depth of the hold being insufficient to provide any resistance to bending as in the conventional steamer.

A series of queen posts fanned upwards from the ship's sides, port and starboard, and iron rods extended from forward to aft along the tops of these posts with stretching screws at intervals, to prevent the hull from hogging when loaded. In spite of their light-weight construction and their general resemblance to a house of cards that a strong wind would cause to collapse, these steamers handled magnificently and showed remarkable speed. An average of 18 knots on a crowded river required powerful engines and great manœuvrability in the ship, and iron nerves laced with immense luck on the part of the pilot.

As one would expect, the engines driving these feather-weight ships were themselves very flimsy and they were built from timber as much as human ingenuity could contrive. They were in a sense house engines, real carpenters' engines, built into and forming part of the ship's fabric and not as separate entities mounted on a bedplate and placed bodily into position. In fact, a conventional paddle engine of the period with its box-type bedplate and massive entablature, iron columns and ponderous cylinder and running gear would be far too heavy for such a hull, it would simply fall through the bottom.

The main characteristic of the Western River steamboat engine, apart from its wooden structure, was that it ran non-condensing and consequently worked at much higher initial pressures than its purely marine counterpart. This enabled the machinery pressure parts—boilers and cylinders—to be made fairly small in diameter and consequently of reasonably thin sections. The weight was further reduced by dispensing with the jet condenser, air pump and associated piping and entrained cooling water, and exhausting straight to atmosphere. The machinery weight was reduced still further, in fact pared down to the barest minimum, by working both boilers and engines at their utmost capacity with a complete disregard for elementary safety. This frontier philosophy of living recklessly and of building and driving ships without any margin for contingencies, misalignment, hot bearings, wear and tear or silting up of a boiler, was responsible for most of the accidents that occurred on the rivers during the nineteenth century.

It was not always the builders' fault, however. Iron was very scarce in the Middle West and the Deep South in those days and had to be spread out very thinly to go round; it all had to be imported from the Atlantic coast cities. Nothing was made of iron that could possibly be made of wood, and even the connecting rod, the pitman, was built up in cruciform section of three pieces of timber with iron strapheads at each end, reinforced by an iron truss bar along its length.

Western River steamers fell naturally into two categories: side-wheelers and stern-wheelers, and I must upset the commonly held notion that the Mississippi steamboat was invariably a stern-wheeler or 'wheelbarrow', the proportion of side paddles to stern-wheelers throughout the river complex being roughly 60 to 40%.

Stern-wheelers were more popular on the upper reaches and tributaries where side wheels might foul the banks, but the mainline freighters on the wider stretches of the river and the crack passenger steamers were usually fitted with side paddles.

The engines driving both forms of paddle-wheels were similar in type, what early Victorian engineers called the Gorgon and present-day marine engineers the 'winch' type, i.e. a straightforward piston-crosshead-connecting rod-crank direct engine; a long-stroke, light-weight engine placed on the working or weather deck. In the case of stern-wheelers, the paddle drum was supported on two immense timbers projecting aft to clear the rudder, a design which gave rise to the familiar wheelbarrow comparison, and was driven by a two-cylinder engine direct on to the crankshaft with cranks at right angles.

In the case of side-wheelers, each paddle-wheel was driven separately by its own single-cylinder engine placed alongside it, there being no connection between the port and starboard engines. This isolation of each engine the one from the other allowed the paddle-wheels to be driven independently ahead or astern. In many deep-sea paddle steamers with disconnecting clutches one wheel could be rotated ahead by the crankshaft while the other idled, but the crankshaft could not drive one wheel ahead and the other astern. In the Western River steamboat the engines were possibly unique in that they could manœuvre the paddle-wheels in opposite directions.

In each case, therefore, whether side- or stern-wheeler, the engine was placed horizontally at deck-level, the cylinder and guides being fitted to heavy wooden bearers supported on cribwork built up from a side keelson. The cylinder casting and guide plates were not, as one would expect, bolted on to a common baseplate itself fastened to the wooden bearers; they were separately bolted direct to the wooden bed by coach screws, the mass of timber forming the base was considered sufficient to maintain the alignment. The crankshaft or, rightly, the paddle shaft was made of wrought iron, with an overhung crankpin in a cast-iron web, the web sometimes being reinforced by a wrought-iron strap. This unequivocal way of building propelling machinery, the millwright's method, resulted in an engine crude in appearance and inefficient in action, but it worked exceedingly well, and it was very cheap.

The steam cylinder was a simple cast-iron barrel with four valve chests cast on, separate steam and exhaust chests at each end, and possibly the most noteworthy feature of this engine was the simple, spidery valve gear compared with which the

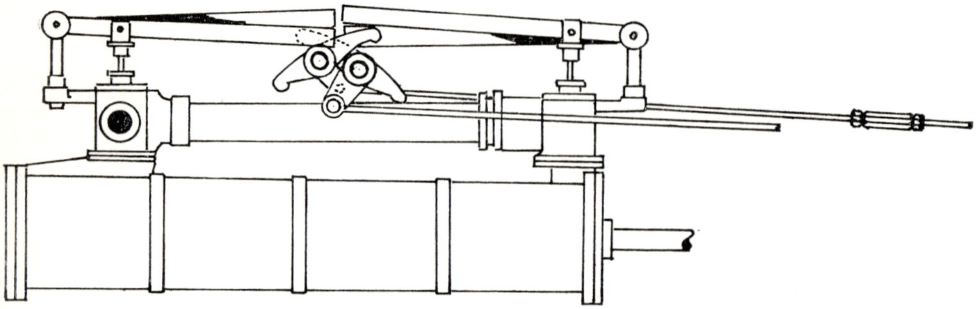

39 Western River steamboat—engine valve gear

contemporary valve motion fitted in deep-sea steamers appeared awkward and heavy in the extreme. The valve lids in the steam and exhaust chests were either of the double-beat type, i.e. double-seated, or what were known as relief valves, that is they were balanced with a small pilot valve let into the main lid, to equalise the pressure above and below the valves. Each valve was lifted by its own fulcrum lever hinged at the end of the cylinder, while the free ends of the levers were lifted in their correct timing by half-moon-shaped wipers on rocking shafts operated by the valve gear from the crankshaft. Small leather pads were placed between wipers and levers to deaden any clicks and the engine was remarkably smooth and mechanically quiet. The only sound in the earlier engines was the whistling in the exhaust valves due to wire-drawing, as both steam and exhaust valves were made of the same bore.

The valve motion was given by a pair of cams mounted on the crankshaft, the one a full-stroke cam to work the exhaust valves, the other a half-stroke cam, sometimes variable, to control the steam valves. These cams were given a profile such that they would open and shut their valves as quickly as possible, and they worked in a large square iron frame which did duty as the eccentric sheave.

The steam for these engines was provided by light-weight horizontal or locomotive boilers working at from 100 lb to 150 lb, and these were invariably placed at the forward end of the working deck, surrounded by stacked firewood collected at each stopping place. They were of the order of 3 feet in diameter by 30 feet long, with a shell thickness of between $\frac{5}{16}$ inches and $\frac{3}{8}$ inches, and normally carried no water-level gauges, only a series of steam and water test cocks. They used raw water from the Mississippi and her tributaries, water rich in acids and alluvial silt in suspension. Reports on boiler repairs in these vessels often referred to the mud being dug out of the boiler bottom with a spade.

To a marine engineer of a later age, accustomed to spending his working life in artificial light or no light at all, and at a temperature well above the comfort level, the placing of a ship's machinery on the weather deck, in a sort of Dutch barn exposed to wind and weather seems contrary to his sense of the fitness of things: a ship with all her machinery laid bare to the vulgar gaze would somehow appear naked or incomplete. And yet how pleasant to spend one's watch on deck in the sunlight and fresh air, with plenty of company with which to pass the time.

The Americans in the Middle West and in the lower reaches of the Mississippi soon learned to make the most of steam by using tugs or towboats to propel a whole fleet of lighters. (In this context a tugboat *pulls*, a towboat *pushes*.)

These lighters were clustered about the towboat, all lashed together, and the fleet usually carried its own fuelling barges with it. One case mentioned a stern-wheeler surrounded by no less than fifty-six coal lighters and two fuel boats, the length of this floating island being 1132 feet and the width 312 feet—that is, an area of nearly 6 acres. In all, this fleet carried 53,200 tons of fuel, the whole propelled by a single light-weight steamboat.

In the course of time compound expansion came to the Western Rivers, and the newer ships, built later in the century, were provided with tandem compound machinery, the HP and LP cylinders being sometimes joined together, the LP piston having two piston rods which engaged in the crosshead yoke together with the HP rod and the steam expanding direct from the HP cylinder into the LP. One of these compound-engined vessels had a stern-wheel 29 feet long by 29 feet in diameter working on a shaft 35 feet long, 15 inches in diameter and weighing 25 tons. The cylinders, 19 and 44 inches bore by 9 feet stroke, worked a 40-foot-long connecting rod. Steam at 140 lb was supplied by six externally fired locomotive boilers.

To complete the picture, here is the description of a Western River steamboat leaving her berth, as given by an eminent English Victorian engineer:

> 'She was steered close inshore amongst stones and stumps of trees, where she lay for some hours to take in goods; the additional weight increased her draught and caused her to heel a good deal, and when her engines were put in motion, she actually *crawled* into deep water on her paddle wheels. The steam had been got up to an enormous pressure to enable her to get off, and the volume of steam discharged from the escapement pipe (i.e. atmospheric exhaust) at every half stroke of the piston made a sharp sound, almost like the discharge of fire arms, while every timber in the vessel seemed to tremble. The numerous

explosions of boilers on the Mississippi cannot excite astonishment amid such provocations as that here described.'

The Eastern River steamboats had many characteristics and peculiarities in common with their Western counterparts: very shallow draught, narrow gutted hulls with wide guards on their working decks and with high many-decked superstructures. External strengthening was basically the same, and to crown all, the engines in both types were mainly built of timber. The resemblance was little more than superficial, however; there was a vast difference in design of hull, engines and boilers between the Eastern and Western genera.

The Eastern River steamboats, in their heyday round about the 1850's, were confined to a fairly restricted region on the Atlantic seaboard within hail of New York. They plied on the Hudson between New York and Albany, the jumping-off point for travellers to Canada, the Great Lakes and the great northwest, a service, incidentally, originated by Robert Fulton in 1807 with his *Clermont*, the first successful steamer in the world. Another line ran between New York and the ports in Long Island Sound which in turn were connected by rail with the New England states.

The Hudson River, normally a deep river, was very shallow near Albany, there being a limiting draught of eight feet, and all river traffic to that city had to conform to these draught conditions. Also, on the Long Island Sound run, although the steamers were in fact sailing in salt water, the channels were very shallow, with a limiting draught of nine feet. The same type of steamer covered both services, large passenger-cargo ferry boats of up to 350 feet in length, with finely moulded lines and towering superstructure, and capable of housing and sleeping up to 1000 people on the overnight passage on a scale completely unknown in England at that time.

The hull was of shallow depth and very light scantlings, with a long hold extending for almost full length which served as the second-class cabin-dormitory. The working deck, as in the case of the Mississippi steamers, was extended well beyond the moulded breadth to form the guards, which were flush with the outer face of the paddle boxes (wheelhouses in USA) and contracted at the stern, but still continued round it in the form of an overhang. All superstructure decks were made of the same breadth, thus presenting a remarkably high and wide structure on a fairly narrow beam. The main or working deck served as cargo space and contained the various offices and crew accommodation, while the first-class passengers were accommodated in the various upper decks. From contemporary descriptions, saloons and sleeping cabins were most luxurious, and the catering, as one would expect, was immeasurably superior to the seafaring passenger's hard tack and hash.

Now a ship of this type, with light-weight platform deck piled high above deck like a house of cards, and with very heavy concentrated loads in way of the boilers, engine and paddle bearings, would fall apart in the lightest swell without some sort of structural support, and here again we find a certain resemblance to the Western steamboat. In the latter a series of queen posts supported a number of iron rods attached to the stem, stern and other portions of the hull requiring extra strength and this complex formed the 'hog chains' whose iron links could be tightened or eased at will. The Eastern steamer used a more sophisticated technique and substituted the hog frame for the chains. A massive wooden arched truss was fitted on the port and starboard sides, tapering down to deck-level forward and aft, but rising well above the superstructure outline amidships to take care of all longitudinal bending, while masts were located at strategic positions on centre and side keelsons, with iron guys stretched down to the heaviest sections of the ship. These masts served only to distribute the loads, they served no other purpose, and, as usual, each guy was equipped with a bottle screw to harden or ease the tension. This system of trusses and guys was not infallible, however, as photographs of the Albany ferry *Albany*, show a pronounced sag amidships in way of her engine room.

The boilers, working at about 35 lb steam pressure, were placed on the working-deck guards, port and starboard, for several reasons, the principal one given being that if they exploded they would 'take off' outwardly with the minimum of damage. Possibly the real object in so placing the boilers was the ease of renewal; these high-duty boilers had a very short life and in some cases had to be renewed within two or three years.

The Eastern River steamers were in the main propelled by side paddles, and their machinery was based on James Watt's rotative beam engine with very little change in style or detail. Originally designed for the Hoboken Ferry in 1822 by Robert L. Stevens, who helped his father John build the *Phoenix* in 1807, this engine remained substantially unchanged throughout the whole of the nineteenth century and was still to be seen in New York Harbour in the 1930's: Watt's beam engine side by side with high-pressure turbines and modern diesels and still holding its own. What a magnificent monument to the inventors Watt and Stevens!

The engine room formed a deep well on the centre line of the ship, and its casings extended through the hold, working-deck spaces and the superstructures up into the skylight, with observation windows at all decks. The engine itself consisted of a tall wooden gallows frame, like the letter A, consisting of a pair

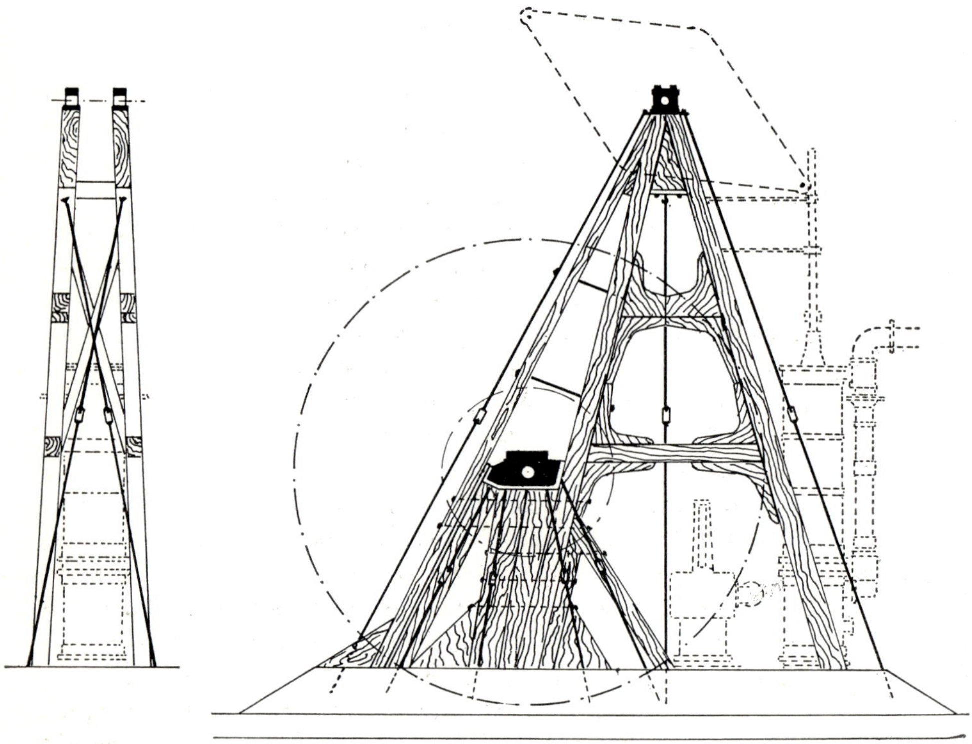

40 Eastern River steamboat engine, with timber gallows frame (1850)

of trestles resting on the main keelson with the single cylinder between its forward feet, the beam fulcrum at its apex and the paddle-shaft bearings supported on the after legs. This wooden frame was solidly built and reinforced at all strategic points by guys and stretching screws to the keelson. When the engine was running these guys could be adjusted to ease a hot bearing or take up wear in the pillow blocks.

The single cast-iron cylinder stood vertically upon the jet condenser and the piston rod worked a T-shaped crosshead, whose ends worked in guides mounted on the cylinder top, braced on to the wooden trestles. From this crosshead a pair of short connecting links laid on to the forward end of the walking beam (literally working beam), a lozenge-shaped structure pivoted about the top of the gallows, which in turn communicated the movement to the crank through the main connecting rod. Because of its length, the connecting rod was trussed in the fore-and-aft direction.

41 Eastern River steamboat—view of working beam (1898)

The merit of this marine engine lay in its slow rotation and immensely long stroke, which varied according to the size and speed of the steamer, but on occasion could reach up to 15 feet, demanding a cylinder casting 20 feet long. This long stroke in turn required an unusual type of paddle shaft. The main shafting extended from port to starboard, with a wheel hub at each end and a single crank in the middle. The paddle-wheel wooden spokes were driven into sockets all round the three hub discs keyed to each shaft end, there being an outer bearing on the sponson or guard. Flat paddle boards or buckets were then clamped to the spokes with U bolts, a method which allowed broken boards to be readily renewed.

The crank webs were shrunk on to the shaft and locked into position with a screwed dowel pin. The crankpin was similar in principle to the crankpins of ocean-going paddle engines in being self-aligning. It consisted of a parallel portion for the connecting-rod big-end bearing, with a tapered portion at each end to engage in the webs. The end carried by the 'driving web' was hammered in and locked with a cotter, while the other end ran loosely in the eye of the 'driven web'. This latter taper had four rounded brass keys let into keyways on the taper, held in place by an end locking plate. This plate was adjusted to maintain the brass keys slightly proud of the iron taper surface to take the drive.

The walking beam was Stevens' development of Watt's cast-iron beam, and consisted of a cast-iron skeleton with four sockets for the main working bearings: crosshead links, fulcrum, connecting rod-top end and air-pump drive, the whole reinforced by a heavy wrought-iron lozenge-shaped truss, whose major and minor axes were in the proportion of about 2 to 1.

The engine valve gear, another of Stevens' patents and similar in principle to that of the Mississippi steamboat's engine, had a steam and an exhaust valve at each end of the cylinder, the valves being usually of the balanced, double-seated poppet type, placed vertically, with outside spindles worked by spring-loaded lifting rods. Each of the four lifting rods, placed side by side between the large flared steam and exhaust chests, carried three brackets or toes spaced at intervals along their lengths: the main toes, the emergency and hand operated ones. The lifting rods, and hence the steam and exhaust valves, were operated by wipers carried on two rocking shafts, one for the steam admission, the other for the exhaust. These two rocking shafts derived their motion from a pair of fixed eccentrics on the crankshaft, one on each side of the crank, through adjustable bell-cranks. There were no astern eccentrics as both the steam and exhaust eccentrics, although given some lead, were roughly in line with the crank, and enabled the engine to run astern with some slight loss of efficiency. The eccentric rod drive to the rocking shaft was by a hook over the bell crankpin, and to reverse the engine this hook was lifted clear by a foot pedal and the steam and exhaust valves worked by hand in the required sequence. As soon as the engine started up in the opposite direction, the foot pedal was released allowing the eccentric rod to click into position. This hand gear, the tripping shaft, was a solid shaft placed near the bottom of the cylinder, at floor-level, having a cam below the starting toe on each valve lifting rod, all four cams moving together, the top steam and bottom exhaust working simultaneously and vice versa. With the eccentrics disengaged,

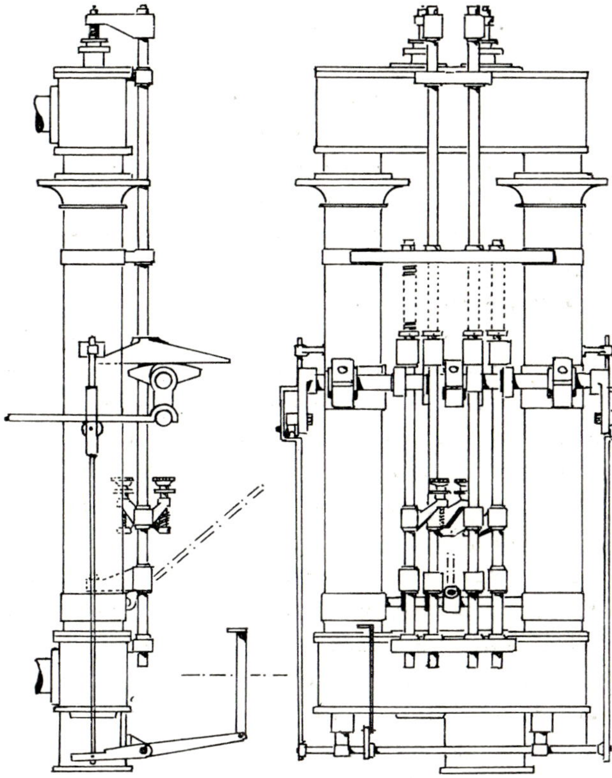

42 Eastern River steamboat—valve gear (1870)

steam could be admitted to either end of the cylinder by rotating this hand-gear shaft with the aid of a portable starting bar.

A third series of toes, the emergency toes, were placed between the running and starting toes, and were given the name of gags. Should a sudden burst of speed be required for a short spell, these gags allowed each steam-valve lifting rod to be worked by its opposite end exhaust-valve wiper, giving the valve full opening throughout the stroke. Alternatively, the hand-operating gear could be used to keep the steam valves open for a longer period of stroke.

The valve gear just described was that of the more sophisticated models built in the 1860's and developed from Francis B. Stevens' patents of the 1840's. The earlier gears, those introduced by John Stevens in 1817, were extremely simple and consisted of a single eccentric keyed to the crankshaft working a single-rocker shaft mounting four fixed wipers, one for each steam and exhaust valve at top and bottom of the cylinder. In this valve motion the steam admission for one end and the exhaust at the other end were open simultaneously for full stroke and the engine worked non-expanding. A later refinement was introduced by fitting the cam-board cut off, a set of cams of varying profile joined together and working in an iron frame rather like an eccentric in a loose-fitting strap. An eccentric rod from this strap closed and opened a throttle valve in the steam pipe in such a way that even with the cylinder steam valve open at, say, half stroke no steam would be admitted. By pushing the frame from cam to cam with successively shorter travel, a variable cut off could be obtained with consequent economy. This butterfly cut-off device was better than nothing but it still was not very efficient, as quite a lot of the steam expanded in the steam pipe and not in the cylinder. With the introduction of the double-rocker shaft and the separation of the steam and exhaust valve movement, a certain amount of lead could be given to the exhaust valve thus easing the reversal strains in the running gear.

In the description of the Mississippi steamer, a Victorian engineer's views were quoted on the handling of the engines leaving the shore in very shallow water, and it may be deduced from that incident that he held a pretty poor opinion of the indigenous marine engine and its builders. Here is the same man writing about the Eastern River paddle-boat engines:

> 'The American steamboat engine has long been a subject of wonder to the English engineer. It is ugly, straggling and inconvenient looking; its incompactness, want of snugness and economy of room make it the reverse of everything we think good in a steamboat engine. It certainly made the same impression on me that it has done on all my countrymen, but it was at first sight, and at first sight only. Daily the unfavourable impression became mitigated by familiarity, and after a careful study of its details and qualities, I do not think it possible to design an engine more admirably fit for its use and purpose under the circumstances where it is applied.
>
> 'In this country there is not a single engine which can be said to be entirely English and pre-eminently suited to any one purpose. The American walking-beam engine on the other hand is universal in the States and acknowledged to be the best suited to their Eastern River navigation. I think it will be admitted, therefore, that the permanence of this kind of engine in the American steamers must be held as prima facie proof of its excellence, and that it is entitled to our respectful consideration and likely to reward our careful study.
>
> 'I have examined its structure in the best workshops of America and have watched its practical working in their best steamboats. I have satisfied myself

that it is cheaper in construction, lighter in weight, more economical in management, less costly in repair, more durable and better suited for high speed than any of our engines could be. I think that for the navigation of large rivers, like those of China and India, it might be adopted with great advantage, and many of its details indeed might be adopted with great advantage in any engine. One great advantage which the walking-beam engine possesses is that of being the only one which imposes no restraint on the engines with regard to length of stroke or diameter of wheel. The beam engine takes up but little additional room in the ship in proportion to the increase in length of stroke, but only raises the walking beam higher above the vessel.

'It will be seen that the shallowness of the water determines one material point in the structure of the engine. It becomes impossible to get a long stroke directly under the shaft, the position in which the English engineers have shown themselves so anxious to place it. Driven from below the shaft from want of height, the Americans have placed their cylinder forward of the crank, and with the centre of the cylinder on a level with the shaft. This necessarily throws the walking beam to more than four times the length of the crank above the level of the shaft which, with a 15 foot stroke, is equal to 30 ft. This also determines four times the length of the connecting rod, and four times its length for that of the walking beam. These parts are therefore in tolerable working proportion, and they necessarily determine the arrangement of the minor parts of the engine.'

The engine rooms for these large engines were very well equipped and finished, and a contemporary journal describes the s/s *Alida*'s machinery space bulkheads as being lined with mirrors in which all the movements of the highly finished machinery were reflected. A far cry from the dirty grey bulkheads, stone-coloured engine and greasy wooden platform of the nineteenth-century European tramp.

The stokeholds too, in these luxury ferry boats, were equipped with machinery far in advance of contemporary sea-going ships. The closed stokehold was introduced in 1845 by another member of that great engineering family, Edwin C. Stevens, and by the last quarter of the century was quite common on these fast ferries.

This American beam engine, born in 1822, came of age in the 1840's and lived to a ripe old age; engines with timber gallows were still being built in 1899 and there was very little difference in general appearance in the type throughout the century. Certain modifications were made in the light of progress, but they were not immediately obvious to the layman, and probably the only one of any consequence to the historian was in the material for the main framework. In the

43 Eastern River steamboat—cut-away view of engine room (1898)

crack steamers with high-powered engines, steel box frames took the place of the wooden trestles, mainly because of the enormous weights and pressures involved and the inherent weakness of wooden joints in tension. Other changes took place, of course: conversion of jet condensing to the surface condenser and, for the choppy waters of the Long Island Sound, iron feathering paddles in place of the large radial paddle-wheels on the smoother rivers. Finally, a radical change in one or two of the largest ships, where compound cylinders were fitted, the LP piston driving the walking beam at its end and the HP, of shorter stroke, at some point nearer the fulcrum. But it must be borne in mind that in spite of some compound surface-condensing steel gallows engines, this native American marine beam engine was essentially a single-cylinder, jet-condensing wooden engine driving large wooden paddle-wheels.

Although peculiar to the American Eastern Rivers, the walking-beam engine must not be considered as exclusive to them, it was also to be found on the Great Lakes and as a genuine ocean-going marine engine. In the latter cases, however, it was not so noticeable: it was not housed in a Dutch-barn type of steamer with the beam jerking to and fro well above the engine-room skylight, but in a conventional type of ship where it would not be seen from outside.

Indeed the s/s *Vanderbilt*, a wooden paddle steamer built in 1856 on the New York–Southampton–Le Havre run, and the biggest steamer in the world at that time, was propelled by two such beam engines, and was unique in that respect. Each engine drove its own paddle-wheel separately, the cylinder being 90 inches in diameter by 12 feet stroke, taking steam at 18 lb. Having independent paddles it must have been one of the very few deep-sea paddle steamers that could turn on her centre with one wheel going ahead, the other astern.

This engine was also fitted on some of the Pacific Mail steamers trading between San Francisco and Panama and was still to be met with on the steamy coasts of Central America. The disadvantage with these coasters, common to all paddle-wheel cargo ships, was that when loaded their wheels were too deeply immersed and tended to choke. In fact, in some cases the engine speed slowed down to some four or five revolutions and the engineer had to stand by with his starting lever handy to help the engine over the centres with the hand gear.

Probably the finest deep-sea vessel fitted with a walking beam engine was the s/s *China* on the San Francisco–Yokohama–Hong Kong run. Built in 1866 with a wooden hull, she was barque-rigged with the smoke-stack well forward. Her engine had a steam cylinder 105 inches bore by 12 feet stroke while the piston rod was 11 inches in diameter, the crankpin 14 inches and the crankshaft 22 inches. To match up to these scantlings, the main steam pipe was 28 inches bore.

8 Steam turbines

Earlier on in this survey of the marine engine it was pointed out how steam, as a gas, was unique in that it could provide power in several different ways: by creating a vacuum below a piston and letting the atmospheric pressure do the work; by using its elastic nature to push a piston to and fro inside a cylinder; or by acquiring a very high speed when its pressure was released through a set of nozzles. The first characteristic was demonstrated in the atmospheric engine, the second in the various types of piston engines and the third in turbines which now come under review.

The turbine was developed rather than invented, it had no inventor. Hero of Alexandria, who was the first to demonstrate the possibility of steam to produce motion in his aeolipile in 130 BC, did not invent the turbine, it existed all around him throughout the Nile Delta region in the form of a crude water-wheel to lift water for irrigation. The water-wheel, in fact, was the first turbine and quite possibly it was developed independently in a number of areas in southern Asia before arriving in the Mediterranean Basin a few centuries before the Christian era. The water-wheel grew more and more sophisticated and efficient with the passage of the centuries, from the old original Nora wheel used in China through the rough flat paddle-wheels of the Middle Ages and the Renaissance to the water turbines of the late nineteenth century. Hero's little toy, a spherical boiler rotated at high speed on trunnions by two reaction arms opposite each other, became the Barker water-mill in the eighteenth century, greatly improved in the nineteenth by Whitelaw and perfected by Thomson in his Vortex water turbine round the middle of last century. Other water turbines come easily to mind: the Francis, the Girard and the Pelton machines were extremely efficient and they worked either on the pure reaction or combined impulse-reaction principle. In other words, turbines existed for a long while before they were applied to marine propulsion.

In steam turbines, the kinetic energy of the steam is used to drive a series of blades in the periphery of a rotor which is made to revolve at high speed, there being two fundamentally different ways of using the steam. In the impulse turbine, the steam acquires a high velocity by expanding through a series of stationary nozzles to impinge against a row of buckets on the circumference of a wheel. In some small turbines, notably the De Laval type, the whole of this kinetic energy can be absorbed in one wheel, i.e. the steam can be expanded from boiler to condenser pressure directly in one single stage, but this usually results in a very high rotational speed. Maximum efficiency requires the total energy of the steam to be given up to the rotor, i.e. that the steam speed at exhaust shall be zero. This entails a bucket or blade speed of half the steam speed, but as the speed of steam, in expanding from, say, 180 lb, down to condenser pressure can approach 4000 feet per *second*, some provision has to be made to reduce the rotor speed by utilising this kinetic energy in stages. This can be done in two different ways: by pressure compounding and by velocity compounding. In pressure compounding, instead of the steam expansion taking place in one stage, i.e. one set of nozzles and one set of moving blades, the total expansion is broken up into a number of stages, each consisting of a set of nozzles and a row of moving blades, the steam expanding by steps down the pressure scale and the nozzles and their respective blading increasing in size to suit the greater volume. The turbine blades or buckets (usually blades are for reaction turbines, buckets for impulse type) are mounted on separate discs threaded and keyed on to the rotor spindle, with a stationary steamtight diaphragm between each wheel carrying the nozzles, the pressure drop taking place in the nozzles. This design of impulse turbine was mainly developed by Zoelly, using up to ten-stage expansions, and by Rateau using up to thirty stages. Velocity compounding, on the other hand, as used in the Curtis turbine, reduced the blade speed and hence the rotational speed incorporating a number of rows of moving buckets running between rows of fixed blades mounted on the casing, the steam sequence being: expansion through a row of nozzles on to the first set of moving blades, the emerging steam being redirected through the first row of fixed blades on to the second row of moving blades, being again redirected on to the third row. In general, in this velocity compound turbine, the two or three rows of moving blades are mounted onto the rim of a single wheel for each stage, the compounding being made up of a few such stages.

Impulse turbines, therefore, are characterised by relatively short high-speed rotors composed of a series of disc wheels keyed on to a spindle, and by the arrangement of their high-pressure steam nozzles whereby, at reduced propeller speed, certain nozzle boxes can be isolated by a separate stop valve, thus reducing the quality of steam to the engine without affecting its pressure in the initial stages.

In the reaction turbine the steam is expanded continuously in either moving or fixed blading throughout, from boiler to condenser pressure, the blade length and the rotor diameter increasing in size to accommodate the greater volume of the expanding steam. In this type the rotor is usually a drum with row upon row of reaction blading round its periphery, while corresponding rows of similar blading are inserted in the turbine casing, each row of fixed blades alternating with one row of moving blades, such that when expanding through them the steam reacts against them to cause rotation of the drum. This reaction of the steam against the blades, forming as it does a pressure drop across each row of moving blades, produces an axial thrust that is balanced by a dummy piston which also acts as a large-diameter labyrinth gland. The complete reaction turbine usually consists of several units, the small but very robust HP rotor and casing and the lighter and much bigger LP turbine. Sometimes an IP rotor is interposed and on occasion a turbine installation can be quadruple expansion.

Now during the last two decades of the nineteenth century many inventors had experimented with steam turbines of different types, De Laval, Curtis, Zoelly, Rateau and others, but it took the genius of Sir Charles Parsons to show how a steam turbine could be made to drive a ship and he chose a very dramatic way to prove it, at the Spithead Naval Review in 1897.

Parsons built his first turbo-generator in 1884, a 8½ kW machine running at 18,000 rpm and by 1888 he had several 75 kW electrical sets running in power stations. However, he had always had the idea of driving a ship by means of the turbine at the back of his mind, and in 1894 he started work on the construction of the *Turbinia*, the first-ever turbine steamer.

The machinery for his ship consisted of a single-casing 2000 SHP radial-flow turbine running at 2000 rpm on a single screw, and trials were held towards the end of 1894. They proved to be very disappointing, the top speed being below 20 knots, and the whole propulsion system had to be redesigned. The single-compound turbine was removed and replaced by three separate-compound turbines coupled to three screwshafts working in series, there being the HP and the astern turbines working the central shaft, and the IP and LP turbines working the two wing shafts, all manœuvring being by the central turbine only. The total weight of the engines was the same as before, the power being increased to 2300 SHP at 2200 rpm, and each shaft was fitted with three screws.

Trials with the new engines were held in December 1896 and gave a speed of 29·6 knots. After experimenting with different shapes of propeller the speed was gradually increased on successive trials to 32·5 knots and on her unscheduled demonstration run at the Spithead review she claimed to have made a speed of 34·5 knots. To complete the picture, the *Turbinia*'s engines took steam from a single water-tube boiler working at 170 lb under forced draught conditions, the fan being driven from the forward end of the HP (central) turbine and the condenser circulating water being supplied by an under-keel scoop.

Now Parson's turbine, due to its simplicity and efficiency, would have made its way in the world of shipping by merit alone, it would have sold itself, as it were, but there can be no doubt whatever that the sheer audacity of the *Turbinia*'s dramatic career past the assembled navies of the world at Spithead earned her the publicity which no amount of inspired salesmanship could command. She captured the attention of the civilised world, and within months estimates for naval and merchant turbine engines were being requested.

The first Admiralty vessel to be equipped with turbine propulsion was the torpedo-boat destroyer *Viper*, ordered in 1898 and completed in 1900, with a total of 11,000 SHP on quadruple shafts, there being two propellers to each shaft. In this case, the engines on the port and starboard sides were duplicated, the HP turbine being on the outer shaft and the LP and astern turbines on the inner. These four shafts gave her speeds of 35·5 knots ahead and 15 knots astern, while a cruising turbine of modest power was fitted at the forward end of the two inner shafts. The *Viper* was completed in 1900 and appeared to be most successful, although unfortunately she was wrecked in a fog off Alderney in 1901 and was therefore not given much opportunity to prove her turbines over any length of service.

The first commercial turbine-driven vessel was the Clyde pleasure steamer *King Edward* with a triple-shaft arrangement, the HP engine driving the central shaft carrying two propellers, and a LP engine on each wing shaft equipped with a single propeller. The *King Edward* was an anomaly in being a quadruple screwed steamer with three shafts only, but engineers in those days were troubled with cavitation at the propeller blades due to the very high turbine speeds, and consequently propeller speeds, causing the water to 'gasify' beyond a certain critical speed. Even today, after more than a hundred years of studying propellers, naval architects are very lucky to arrive at the optimum propeller at first try for some of our largest passenger ships.

The *King Edward* was built in 1901 and was an immediate success, cruising at 20 knots without any effort. After stories of her success had circulated for a while, orders were given for two cross-channel steamers, the *Queen* and the *Brighton*, delivered in 1903, both very similar to the Clyde prototype. The *Queen* was a triple-shaft arrangement, the HP turbine driving the central shaft at 700 rpm carrying a single propeller, while a LP and astern turbines drove each wing shaft at 500 rpm, each shaft equipped with two propellers, there being a total of five. The propellers varied in size: the HP being 3 foot 6 inches disc diameter and the four LP-cum-astern propellers 2 foot 3 inches. Full ahead was on all three shafts, astern running on the wing shafts only. Steam was provided by two double-ended and two single-ended Scotch boilers at 150 lb, and at full power this ship could make a comfortable 20 knots. The *Brighton* was a simpler version of the *Queen*: she had the usual triple shafts, but with each driving a single propeller only, the turbine arrangement being similar.

By 1904 there were two Clyde steamers, four cross-channel packets and five colonial coasters driven by turbines and the time had come to venture out into the Western Ocean and test these new engines under realistic heavy-weather service conditions. It must be borne in mind that half the total turbine power is developed in the LP engine, where a good vacuum is essential and fears had been voiced in engineering circles that the turbine might prove to be inefficient in tropical waters.

Though these have since been shown to be groundless, they could not be just ignored by a ship-owner prepared to gamble a fortune on the success of the new engine.

The first turbine steamers on the Atlantic Ferry Service were the *Victorian* and the *Virginian*, passenger ships of some 11,200 tons commissioned in 1904, the one built in Belfast, the other in Glasgow. These sister ships had the by now common arrangement of triple screws on three shafts with the HP on the centre and an LP and astern turbines on the port and starboard wing shafts. The vessels could sail at 17 knots and their shafts rotated at 270–300 rpm. As with the coastal and channel steamers, they manœuvred on the wing shafts as a twin-screw vessel.

The Cunard Steamship Co. were impressed by the performance of the two Allan Line twins and they decided to build a pair of 20,000-ton passenger steamers that would be similar in all respects except in regard to their machinery. The *Carmania* was fitted with triple-screw turbines with the conventional layout: HP on the centre shaft, the LP and astern engines on the wing shafts. The *Caronia* had orthodox twin-screw quadruple-expansion engines, and both ships developed 20,000 SHP at 19 knots, the steam in each case being provided by eight double-ended and five single-ended boilers steaming at 185 lb. The *Carmania*'s engine room was an enormous cavern housing immense turbines running at 180 rpm. The two LP turbines were 14 feet in diameter and 36 feet long and the total weight of her machinery amounted to more than that of her sister ship.

However, the *Carmania*'s performance was superior and from then on all Cunard Atlantic passenger steamers were to be driven by turbines.

Here it might be pointed out that while the *Victorian*, *Virginian* and *Carmania* were the first turbine-driven passenger liners on the Atlantic Ferry, they were not the first turbine steamers to cross the Atlantic; that honour belonged to the steam yacht *Emerald*.

By now, the Parsons marine steam turbine had become well established and some idea of the phenomenal success of this engine will be shown by statistics provided by Sir Charles Parsons himself. In 1903 the marine turbine was represented by some 23,000 SHP for both naval and merchant ships. In 1906 the total had risen to 830,000 SHP and by 1909 it had reached 3,000,000 SHP. This engine had proved itself in the mail boats on the Western Ocean, a very exacting service on the worst seas in the world, and in 1906 the two Cunard giants, *Lusitania* and *Mauretania*, were launched, sister ships of 43,000 tons displacement developing 68,000 SHP on quadruple screws. In these two ships the engines were duplicated, the HP turbines driving the port and starboard wing shafts, the LP and astern turbines driving the inner shafts. Steam at 180 lb was provided by twenty-five boilers, twenty-three of them double-ended. As the *Mauretania* is probably the largest and the best known of these giant direct-drive turbine ships, a more detailed account of her engines is reserved for a later chapter. This vessel was not, however, the last to use the direct-connected turbines, the record for this among the giant transatlantic ferries goes to the *Aquitania*, built in 1914 with a total of 60,000 SHP.

Although the Parsons' reaction-type turbines were used in all of the steamers described above, that is not to say that it was the only type available for propulsion. The steamer *Creole*, built for the Southern Pacific Railroad Co. in Massachussets in 1907, was driven by a pair of Curtis impulse turbines driving directly on to twin screws, each turbine developing 4000 SHP with steam at 225 lb, and several inventors, De Laval, Rateau and Zoelly, had produced workable steam turbines, but Parsons was the first in the field and his engine was already firmly established before any other type became known.

Now it was mentioned earlier in this chapter that steam expanding from boiler to condenser pressure through a single-stage nozzle acquired a speed of some 4000 feet per second, entailing a bucket peripheral speed of 2000 feet per second, i.e. in a single-stage turbine with a disc diameter of 7 inches the rotational speed would amount to 60,000 rpm.

In the early 1900's this speed was of no use to anybody, and the only way to utilise the power of the steam efficiently and usefully was either to use a large number of expansion stages or to increase the diameter of the rotor. In the case of the *Mauretania*, by using a large number of expansion stages on two rotors, the shaft speed was brought down to a modest 190 rpm but this still required a LP turbine (ahead and astern) some 90 feet long and about 16 feet in diameter. The sad fact is that the steam turbine is essentially a high-speed engine while the screw propeller is a slow-speed drive, and although direct drive was practicable in naval and high-speed passenger steamers with fine lines and high propeller speeds it was impossible in the blunt-ended cargo boats with large slow propellers and small engine rooms.

To integrate the steam turbine into the ordinary merchant steamer an additional factor was required: reduction gearing. Not the crude cog wheels such as Brunel used in his *Great Britain*, but a precision gear train capable of transmitting anything up to 50,000 SHP at extremely high rotational speeds, and with an efficiency of from 97 to 98%.

De Laval was apparently the first to use reduction gearing to a turbine, in 1888, followed by an experimental set of gearing in 1892 on a small marine unit. In 1897 Sir Charles Parsons experimented with a 10 HP turbine geared to a twin-screw launch with a 14:1 reduction gear and obtained a speed of 9 knots, and decided to carry on the experiment on a larger scale. In 1909 the directors of the Turbinia Works bought the s/s *Vespasian*, a small tramp steamer of some 4300 tons displacement built in 1887 and equipped with an ordinary triple-expansion engine of $22\frac{1}{4} \times 35 \times 59$ inch cylinder bores by 42 inch stroke, taking steam from two boilers working at 150 lb pressure, and the usual air, circulating, feed and bilge pumps driven off the IP crosshead. This old engine was overhauled and its HP cylinder was rebored, and the vessel made a voyage from the Tyne to Malta with the trial staff on board to obtain the basic propulsion data, speed and consumption. Upon her return to the Tyne, her reciprocating engines were removed and replaced by a set of Parsons' reaction geared turbines, the first ever installed in a deep-sea vessel. The HP and the combined LP and astern turbines each drove a pinion meshing with a bull wheel on the propeller shafting, this wheel being an iron casting with two forged-steel tooth rims shrunk on, $99\frac{1}{2}$ inches PCD with a face width of 24 inches, the teeth being inclined at 20° to the shaft. The pinions were of chrome nickel steel, 5 inches PCD and the gear ratio was 19·9 to 1.

On trials, the turbines gave the *Vespasian* a speed of 10·7 knots on 1095 SHP, considerably more than the speed obtained with her original engines, and with a reduced consumption.

The original auxiliary installation was retained after conversion, the engine-driven pumps being worked from a small crank at the end of the main shafting, while the only additional machinery required were the two independent lubricating pumps.

The *Vespasian* showed how the steam turbine could be used more efficiently for all types of steamer, but especially for the slow freighter and small intermediate vessels. However, it was not the complete answer. As powers and turbine speeds increased, the reduction-gear ratio had to be increased in like proportion to keep the propeller revolutions within practical limits. This required either a smaller pinion diameter or a larger bull wheel, and as there were practical limits to the sizes of these two components, another solution had to be found. It is true that a worm and worm-wheel drive would have solved the gear ratio problem, but at a heavy cost in friction—the logical answer lay in double-reduction gearing, a speed step-down in two stages.

The *Somerset* was the first merchant vessel to be equipped with double-reduction gearing, in 1918, but it took some considerable time before double-stage gears became established, and at one period in their development some ship-owners considered discarding this type of drive and returning to the original concept of single reduction. The difficulty apparently lay not only in the design of the gearing but in the materials and techniques available both during and shortly after World War I. The single-geared turbine arrangement in a large engine room can consist of two, three or four turbines (HP, first IP, second IP and LP) driving on to the main bull wheel in a number of different ways. Thus there can be the simple layout of HP turbine on one side of the wheel with the LP on the other. In a triple-expansion arrangement the IP and the LP turbines can drive the wheel from each side with the HP rotor between them driving on the upper rim. Or, the HP and the LP turbines may be in tandem on one side with the larger LP and astern turbine on the other side. On one such installation, a quadruple-expansion arrangement in an old cargo steamer, the turbines were placed at each corner of the engine room with the main wheel in the middle, the HP and first IP, placed vis-à-vis with a common, fixed, pinion between them driving on to the wheel on the starboard side, with a similar arrangement for the second IP and the LP on the port. In all these cases alignment was in the first degree: a simple pinion-to-wheel problem, with the bearings housed in a heavy cast-iron casing.

In the case of double-reduction gearing the early sets required separate casings for the primary pinions and their wheels, all meshing into the big wheel in its own casing. This demanded very careful design and assembly throughout, with the knowledge that however meticulously installed on board, the gearing might experience alignment troubles once the ship was loaded and labouring in a heavy sea.

By improvement in design and with better materials and finer machining techniques, the modern double-reduction gearing is now as efficient and probably lighter than its single-reduction counterpart, and its acceptance throughout the world has started another tentative step in speed reduction. Some modern turbine engines using very high-pressure, high-temperature steam now have the HP rotor geared down before meshing into the double-reduction gearcasing, a form of triple reduction, in fact.

The common arrangement of double-reduction gearing was to have the turbine drive its primary pinion through a flexible or self-aligning coupling to the primary

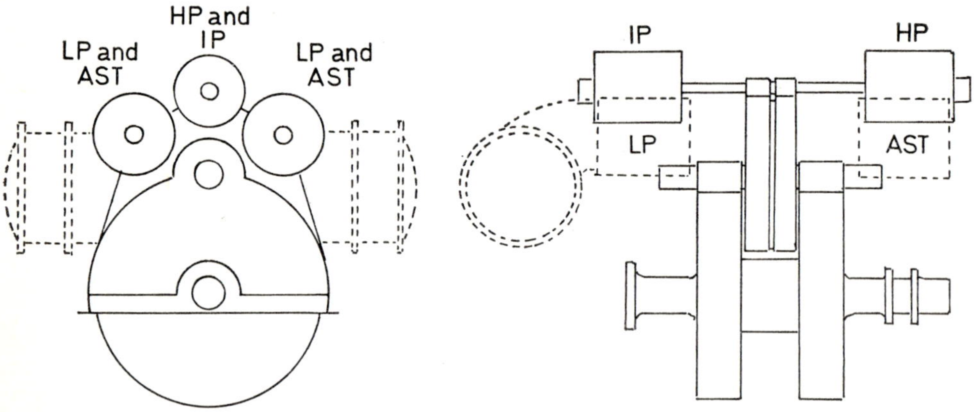

44 Arrangement of Macleod steam turbine and gearing (1949)

wheel, the latter being mounted on a shaft carrying the secondary pinion meshing directly with the bull wheel. This made for a lengthy and sprawling gear assembly, and later gears were sometimes built up on the 'nested' pattern, where the HP and LP primary wheels ran between the two halves of the main wheel, there being a secondary pinion both forward and aft of the primary wheel.

A most unusual assembly of double-reduction-geared turbines was at one time to be seen on the steamer *Marianne Clunies*. This ship, built at Oregon in 1918, was a standard three-island cargo type with the usual triple-expansion steam engine, and she was taken over by the British Government during the late war and renamed *Empire Opossum*. Having sustained serious machinery damage in 1949, she was gutted out and re-engined with a Macleod turbine and her name altered back to *Marianne Clunies*.

Her machinery consisted of a double-reduction gearcase carrying six small impulse turbines: HP, IP, two LP and two reverse rotors, the rotors being coupled vis-à-vis in pairs, each pair with a central pinion meshing on to its individual primary wheel nested into the bull wheel. Thus the HP and the IP turbines drove a common spindle, and each LP was coupled to a separate astern rotor. The six turbine casings were overhung on the gearcase, being bolted thereto on large flanges, and were thus supported at their inner ends only. The rotors were exceptionally light, each turbine consisting of three impulse wheels only, their diameters ranging from 12 inches in the HP and astern turbines to 14 inches in the LP units, while their spindles measured a mere $3\frac{1}{2}$ inches diameter.

These turbines ran at about 10,000 rpm and the total weight of the assembly, turbines, gearing and gearshaft, weighed only 28 tons for 2500 SHP, and with luck this package turbine could have set the trend for a new type of economy engine. However, the installation was not considered satisfactory and the turbines were removed and replaced by a set of twin-geared diesel engines, the name being changed once again to *Ansgaritor*. This old vessel, if she achieved no special distinction, at least holds the record for being propelled, at different stages in her career, by steam reciprocating, turbine and diesel engines.

While Parsons was experimenting with the machinery of the steamer *Vespasian* on the hypothesis that reduction gearing was the alternative to direct-coupled turbines for slow-speed vessels, the Americans were introducing a novel form of reduction unit in the form of electric drive: the turbine would drive an alternator at a speed suitable to both, and the current supplied would be used to drive a large, slow-running motor at a speed convenient to the propeller.

The American-owned bulk freighter *Frieda* was built in 1911 for service between New York and the Gulf ports, and was equipped with just such a machinery arrangement: a turbine direct-coupled to a three-phase fifty-cycle alternator and exciter in tandem at 3000 rpm, with a solidly built motor of 1900 SHP driving the single screw at 84 rpm. Allowing for the usual teething troubles and for the difficulties encountered on the electrical side when electricity was still in its infancy, this installation must have proved successful, even though the paramount argument for electric drive, the elimination of the intermediate shafting, did not apply in the *Frieda*'s case, as her machinery was fitted aft.

In 1912 a US Government collier, the *Jupiter*, was fitted out with turbo-electric machinery on twin screws and she developed 14 knots at 20,000 tons displacement. In this ship one turbo-alternator running at 2000 rpm supplied current at 2200 volts to two synchronous motors, one for each screwshaft, running at 110 rpm, the turbine being a six-stage Curtis machine. Turbo-electric drive was taken up on our side of the Atlantic around 1916, the first vessel to be so engined in UK being the *Wulstey Castle*, built in 1918. This machinery, however, was not retained, as it was subsequently replaced by diesel engines driving through hydraulic couplings. This setback did not worry ship-owners overmuch, as a number of famous steamers have been, or still are, propelled by turbo-electric power: *California, Pennsylvania* and *Virginia* in USA, *Viceroy of India, Strathaird, Strathnaver, Queen* and *Monarch of Bermuda* and even including the largest passenger vessel of her age, the *Normandie*. At the time of writing, the s/s *Canberra* is the largest vessel with this type of machinery, developing up to 45,000 SHP per shaft.

A variant of the direct-coupled marine turbine was the combination drive, an exhaust turbine in association with one or two reciprocating engines, usually arranged as a triple-screw drive, with the reciprocating engines in the wing shafts

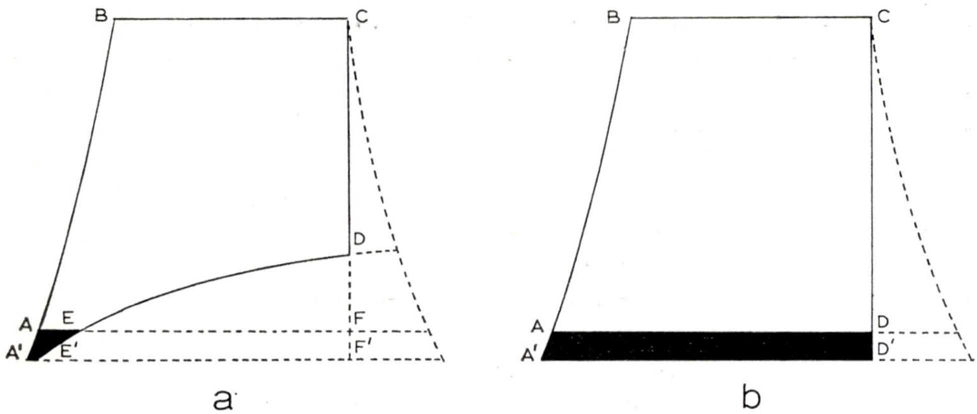

45 Temperature entropy chart showing effect of reducing condenser back pressure in:
(a) a steam reciprocating engine
(b) steam turbine
(*Shaded area represents increase in output*)

and the exhaust turbine driving the centre shaft direct. To appreciate the value of the LP turbine in this combination, it should be realised that a triple-engine cannot work very efficiently with a release pressure less than about 8 lb absolute, as the LP cylinder has to handle and contain the total amount of steam passing through the engine at each stroke. When the steam has been expanded down to near condenser pressure at 26 inches vacuum, its volume is about 200 cubic feet per pound. By increasing the vacuum to 27 inches, the volume reaches some 300 cubic feet and a very much larger LP cylinder would be required to handle it, the cost being out of proportion to the small extra work obtained.

The turbine, on the other hand, has no such limitations in that the low-pressure steam simply rushes through it towards the condenser, it does not have to handle, manœuvre and reverse large volumes of steam through heavy valves or to work heavier pistons. The LP turbine, in short, can use the steam at its lowest pressure, its only limits being brute size and the equipment to provide, and maintain, a very high vacuum. This fact is brought out in the temperature-entropy $T\phi$ chart for the Rankine cycle and shows the theoretical amount of work done per pound of steam, (a) for a reciprocating engine given by the area ABCDEA and (b) for a turbine by the area ABCDA, steam and condenser conditions being identical. The area EDFE accounts for the extra work extracted from the steam by the turbine in continuing its expansion to condenser pressure instead of to the higher release pressure in the reciprocating engine dictated by size considerations. If, now, the condenser pressure were to be reduced by, say, 1 lb per square inch, the line AEF will drop to the lower value A′E′F′. The increase in engine efficiency for the reciprocating engine and the turbine is shown by their additional areas AEE′A′ and ADD′A′, in black.

The first practical test of a combination turbine was carried out on HMS *Velox*, a quadruple-screw destroyer, in 1902. Here, the two inner shafts were driven by a pair of compound reciprocating engines each exhausting into a LP turbine on the same shaft but separated by a clutch from the reciprocator. The wing shafts were driven by separate HP turbines exhausting into the LP on the inner shafts. While cruising, the reciprocators were used, exhausting into their individual LP turbines. At full speed, the reciprocators were declutched and the four screw-shafts were driven by turbine direct, the outer turbines exhausting into the inner LP units.

The first merchant vessel to be fitted with an exhaust turbine was the *Otaki*, built in 1908. The machinery consisted of a pair of triple-expansion reciprocating engines $24\frac{1}{2} \times 39 \times 58 \times 39$ inch stroke, and the exhaust turbine, driving the centre screw, was about 7 feet 6 inches in diameter; it was non-reversing, all manœuvring being carried out on the wing reciprocators. This ship developed 5880 total IHP on trials at a speed of $14\frac{1}{2}$ knots, the average voyage speed being nearer 11 knots. Her two sister ships *Orari* and *Opawa* had conventional steam reciprocating machinery of the same power, yet though the *Otaki*'s machinery weighed 30 tons more, her consumption was some 10–12% better.

Shortly after the *Otaki* went into commission, the first passenger steamer to have an exhaust turbine was launched. The *Laurentic*, 14,500 tons gross, was built in 1908 and had the usual triple-screw arrangement: four-cylinder triple-expansion engines on the wing shafts, with a large non-reversing low-pressure turbine engine on the centre shaft. Manœuvring was done by the reciprocating engines only, each engine having a large section change-over valve at the LP stage whereby the exhaust could be directed to the LP turbine or to the condenser, this valve being under the control of the hydraulic reversing cylinder on each engine.

This triple-screw combination arrangement bid fair to rival the direct-turbine drive for large passenger steamers and became a great favourite with the White Star Line and the Aberdeen White Star Line in the days before the 1914–18 war. The old steam reciprocating engines were still the most popular with ship-owners, but in large power vessels these engines had become huge and unwieldy. Like the dinosaurs of old, they had outgrown their ecology, they had outlived their usefulness. To obtain the power to drive a large transatlantic ferry, the reciprocating engines had to be scaled up to an immense size, and ship-owners were faced with the same problem that had puzzled the owners of the *Great Eastern*. High-pressure steam, i.e. steam over 200 lb, had not yet arrived, and there was a limit to the size and bore of steam cylinders.

It was precisely at this juncture that the combination exhaust turbine showed how additional power could be obtained from the opposite end of the pressure scale, and that the reciprocating engine and the turbine could run, as it were, in double harness. Thus, for a few years, direct-coupled turbines and combination turbines appeared as friendly rivals on the North Atlantic ferry service. The largest examples of the combination exhaust turbine drive were the ships *Olympic*, *Titanic* and *Britannic*, where the total shaft horsepower of nearly 50,000 IHP was developed on triple screws, 15,000 IHP of this being supplied by the exhaust turbine whose initial steam pressure was well below the atmospheric line.

It is odd that in the case of both the *Great Eastern* and the *Olympic*, each the largest ship of her time and each fitted with two distinct types of prime mover, all four types of engine should prove to have been the largest of their kind ever made.

Now while the conventional arrangement of combination turbine consisted of triple screws, this was not invariably the case. The *Rochambeau*, built for the Cie Generale Transatlantique, and the *Reina Victoria-Eugenia*, for the Cia Transatlantica of Spain, were both quadruple-screw vessels, with four-crank triple-expansion reciprocating engines on the inner shafts, and an LP exhaust turbine on each wing shaft.

With the coming of single- and later double-reduction gearing the old-fashioned direct-driven turbine disappeared and with it the combination exhaust turbine, although the steam reciprocating engine was still in favour for the smaller freighters. The thought of all that energy in the LP exhaust being wasted in the condenser still worried inventors, however, and it was not long before the exhaust turbine was revived, this time as an auxiliary turbine drive.

Probably the best known was the Bauer-Wach system, which used a small high-speed turbine, driven by the exhaust from the LP, indirectly coupled to the main propeller shafting through double-reduction gearing incorporating a hydraulic coupling. The turbine pinion meshed with its primary wheel, the body of which formed one half of the fluid coupling. The second half was part of the secondary-pinion shaft meshing direct with the main wheel which was mounted on a torque sleeve on the main shafting. Manœuvring was carried out with the turbine out of gear. To put the turbine into gear (but only with the engine running ahead) a small lever was pulled back on its sector and the lubricating oil pumps filled the fluid coupling, thus bringing the turbine gradually up to speed. When the coupling was filled up, the oil pressure backed up on to the servo-motor piston which changed over the exhaust valve from condenser to turbine. Although apparently complicated in layout, this exhaust turbo system ran very smoothly and gave no trouble at all.

Another type of auxiliary exhaust turbine incorporated the electric drive, the turbine driving a generator through gearing, the current being used to drive a slow-speed motor, either mounted direct on to the intermediate shafting or on a separate seating driving into the main shafting through a chain.

A third type, although not in quite the same category, used an exhaust turbine to drive a turbo-compressor, compressing steam from the HP exhaust and delivering it at a greater pressure and temperature to the IP valve chest.

Finally, to come full circle, a type of combination exhaust machinery appeared in the 1930's composed of either a single- or a double-compound steam engine on one side exhausting into a LP turbine on the other, each unit feeding into a combined gearbox driving a single screw.

Inevitably, with the increase in turbine efficiency through the introduction of gearing, the noise level in the engine room has increased to a degree that would not be tolerated by a landsman. The direct-coupled turbine engine room was a cheerful place to be in, with hardly any noise worth speaking about: the ticking of the dynamo in its corner, the clucking of the main circulator and the wail of the feed pump against the low background humming of the steam in the turbines. The single-reduction gearing gave a pronounced rumbling at close quarters which drowned the sound of the auxiliary machinery, but which was not unpleasant in itself: a level of noise such as would be found in a main street, or in a busy machine shop. But the mad screaming of the double-reduction gearing in a twin-screw steamer must be heard and lived with to be properly appreciated. Stepping into the engine room from the quiet deck outside, the noise assaults the ears and the nerves almost as a physical blow. But that is the price of progress.

9 Some giants in the age of steam

This chapter is really a pause in the narrative of the marine engine saga, a digression in the history of the development of the steam engine from side levers to high-speed geared turbines. By choosing certain engines in outstanding ships over a period of years, I have attempted to recapitulate the story told in preceding chapters. In most cases the engines chosen were the most powerful of their kind, the giants of their respective ages, but before beginning a description of them it might be opportune to define the terms of reference and in this context the dictionary meaning of giant is taken as something of extraordinary strength and stature.

Now ever since steam engines were used to drive ships, early in the nineteenth century, there has always been a largest marine engine somewhere in the world at any given time and in many cases the difference between the largest and the second largest was very small indeed. Again, we who live in the 1970's and deal with ships of 160,000 IHP and over are not unduly impressed by the modest powers developed in the Atlantic greyhounds of a hundred years ago until we pause to consider just how those disparate powers were produced. In a modern Atlantic liner the steam-turbine rotors could be fitted quite easily into the lounge of an average suburban home, while for the passenger steamer of the 1890's it would take a cathedral to house the engines.

However, while the horsepower of an engine will always be its most important attribute, my concern in this chapter is to describe some of the engines which appealed to the marine engineer not so much on account of their strength, as upon their stature, their sheer brute size. By our standards they were not very efficient and not always as powerful as their dimensions might lead us to expect, but they broke all records for magnitude and for bulk, and their like will never be seen again.

The earliest steamers to qualify for this distinction were the three Cunarders *Persia*, *Arabia* and *Scotia*, the last of the great paddle steamers built for the Atlantic ferry service. These vessels were roughly similar in their machinery arrangements with the exception of minor differences in cylinder bore and stroke, the engines being of the side-lever type with twin cylinders, usually referred to as a pair of engines, and these engines were the largest of their kind ever built. Each piston worked its own crank through the intermediary of a crosshead and a pair of side levers, these side levers being joined at their other ends by a crosstail working the connecting rod.

The four side levers were made of cast iron, 6 feet 6 inches deep in the middle and with 25-foot centres. The cylinder bore and stroke varied among the three ships, being 100 inch bore by 10 feet stroke for the *Persia*, 103 inches bore by 9

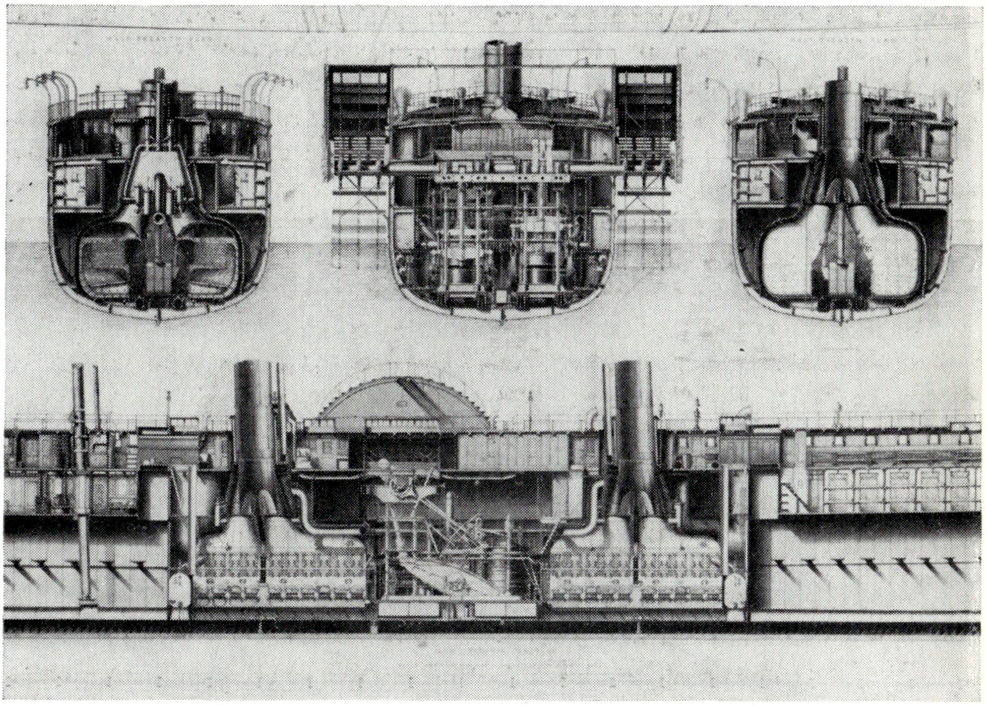

46 Sections through machinery spaces of the paddle steamer *Persia* (1856)

feet stroke in the *Arabia* and 100 inches bore by 12 feet stroke in the case of the *Scotia*. The whole of the framing of each engine was of wrought iron, with cast iron for the shaft entablature. Paddle-wheels were 40 feet in diameter in each ship, 10 feet wide with 3-foot-deep paddles, and with a steam pressure of 22 lb and a rotational speed of 18 rpm, gave a ship speed of about 13 knots. The indicated horsepower of the *Persia*, built in 1856, is given as 3600, while that of the *Scotia*, built in 1862, was 4800, but these figures must not be taken too literally, as the usual practice in those far-off days was to derive the horsepower from a formula based on a nominal steam pressure. Even with steam as low as 22 lb gauge pressure and allowing for the enormous friction in the piston rings and various bearings, a couple of cylinders as large as bedrooms would produce tremendous power at the cranks.

These engines were worked with loose eccentric pulleys, some 4 feet in diameter, driven by stops on the intermediate- (crank) shaft for ahead and astern, each pulley being balanced by a counterweight to keep it from oscillating. To reverse the engine, the eccentric rod was merely lifted off the valve-spindle pin and the engine came to a stop in a revolution or two, working against compression, it being unnecessary (and most inconvenient) to shut off the steam from the boilers. By working the valve spindle in the reverse direction, the piston stroke and hence the crankshaft rotation were reversed, and as soon as the engine commenced to move in the astern direction the lost motion in the eccentric sheave was taken up and the eccentric rod could be dropped on to its valve spindle and work the valve automatically. This operation was carried out by means of the large-diameter 'steering wheel' on the middle platform with half a dozen men climbing round it to a precision timing. Any fumbling resulted in a stalled engine and the probability of serious injury from the kick of the wheel by the valve gear in motion.

Coal consumption in the *Persia* was in the region of 150 tons per day, the bunker capacity being 1400 tons for the eight boilers.

The illustration shows part of the longitudinal section of the *Persia*, engined by Robert Napier, and was painted by David Kirkaldy from the original plans. It is possibly the only engineering drawing ever to have been exhibited at the Royal Academy and for very many years it hung in the waiting room of Kirkaldy's test house in South London before acquiring a permanent home in the National Maritime Museum at Greenwich. Even in the nineteenth century the disappearance of the *Persia* from the Atlantic run made a minor mystery out of a routine business: nobody seemed to know what had happened to her. She was, in fact, broken up in Silvertown on the Thames in 1871. The *Scotia*, the very last of the Cunard Atlantic paddlers, was sold in Birkenhead in 1879, when her paddles and machinery were removed and replaced by twin-screw compound engines. She was bought by the British Telegraph Construction Co. and passed the rest of her active life at Singapore.

Another claimant to the category of giants was Brunel's *Great Eastern*, the biggest ship of the nineteenth century. Launched in 1858, she was exactly twice the length of the *Himalaya*, the largest merchant ship in existence at that time, and the first ship to exceed her in tonnage was the *Lusitania*, built in 1906. She thus held the record among giants for nearly fifty years, in itself a record for shipping. As already explained in a previous chapter, the *Great Eastern* was propelled by both paddles and screw, the engines for which were quite naturally the largest and most powerful of their kind. The screw engines were horizontal reciprocating vis-à-vis type, with four cylinders 84 inches bore by 48 inches stroke, developing about 5000 IHP at 38 rpm and giving the ship a speed of 9 knots on screw alone. Each piston had two rods attached to a horizontal crosshead, the connecting rods from these crossheads driving a single crank for each pair of vis-à-vis cylinders. The pistons on the starboard side worked one connecting rod each, while those on the port side each drove a pair of connecting rods with their crankpin bearings on either side of their opposite number. There were thus two cranks at right angles to each other. The six connecting rods consisted of two long rods threaded through the top-end and crankpin-end bearings and acting as bolts at their extremities. They were in fact built on the same principle as a tramp engine's pump links.

There was a separate condenser and air pump for each cylinder, the air-pump buckets being wrought off the main crossheads. These jet condensers were placed between the main steam cylinders. The slide valves were controlled by ahead and astern eccentric sheaves, and were linked together by a lattice construction common to both vis-à-vis cylinders. This trellis frame was driven by the eccentric rods and link motion. The link motion was reversed by a pair of steam cylinders immediately above the valve links, but could also be worked by a pair of hand wheels, the cylinders on each crank being reversed independently in keeping with the fashion of the times. The slide valves were of gridiron type, and such was their weight that they worked on rollers within the valve casing.

This horizontal engine extended athwartships from bilge to bilge and accessibility to the crankshaft main bearings and running gear was both difficult and dangerous in view of the paucity of gratings and handrails. There was a fore-and-aft tunnel below the engine bedplate, and an athwartships tunnel between the two pairs of jet condensers, but any feeling round and oiling had to be done from above the machinery, and safety depended upon a cool head and a sure footing.

The propeller was not only massive but interesting as being typical of its age. It consisted of a hollow cast-iron boss 8 feet in diameter, made in two halves bolted together and then reinforced by six shrunk-on iron hoops to preserve the integrity of the structure. The four iron blades were bolted on, each by twelve bolts $2\frac{1}{2}$ inches in diameter, and the propeller, 24 feet in diameter and 36 tons in weight, was attached to the screwshaft by a pair of cotters.

While this four-cylinder horizontal engine was a true giant in its own right, it did not make the same impact on the mind as the vertical paddle machinery on the same ship. A vertical engine always seems larger and more important than a horizontal one, and in the case of the *Great Eastern*'s paddle engine, enclosed as it was in an engine room that fitted like a glove, the impression of height was intensified. It was, in fact, 52 feet high, almost the height of a six-storeyed building, and developed 3400 IHP at $10\frac{3}{4}$ rpm, giving a ship's speed of 8 knots on paddles alone.

The paddle machinery consisted of a vertical four-cylinder oscillating engine with cylinders arranged in pairs in Vee formation, each pair working one crank. The cylinders were 74 inches in diameter with a stroke of 14 feet, each cylinder with its piston, rod and valve gear weighing 38 tons. The intermediate shaft consisted of a crankweb and pin at each end and a short complete crankthrow in the middle for working the air pumps. Each of the two main working cranks consisted of a half-crank on the intermediate shaft with its crankpin meshing into a friction strap on a large cast-iron disc keyed to each paddle shaft. This friction strap enabled either wheel to be disengaged at will.

The two jet condensers with their air pumps were placed between the two pairs of steam cylinders, the pumps being driven by long connecting rods from the central crankthrow in conformity with oscillating-engine practice.

Originally, the paddle-wheels were 56 feet in diameter, later reduced to 50 feet after storm damage.

Steam for both paddle and screw machinery was supplied by ten tubular boilers at 25 lb, and the main steam pipe in way of the two engine rooms was 3 feet 9 inches in diameter, probably the only instance in marine history where the main steam pipes could be examined internally by a surveyor crawling about inside them.

The intermediate shaft was the most difficult part of this engine to make, and it was produced only after two abortive attempts by the forgemasters Fulton & Neilson, of Glasgow, the price working out at £100 per ton, a ruinous figure for those days. When it is considered that the as-forged shaft consisted only of two plain journals with a set-up crankthrow between them one might be pardoned for wondering wherein lay the forging difficulty. This lay in setting up a 3-foot crankthrow in a solid 24-inch-diameter shaft, an almost impossible undertaking with the rudimentary steam hammers of the early 'fifties. The crankwebs for the steam-driving cylinders were shrunk and keyed on. Some figures for weights might suggest the size of this huge paddle crankshaft:

Two paddle shafts, each	30 tons
One intermediate shaft	31 tons
Two cast iron discs, each	9·5 tons
Two crankwebs, each	7·5 tons
a total of 125 rotating tons.	

This monstrous engine weighed about 600 tons without the boilers, and developed little more than 3400 IHP at full power: today a high-speed diesel engine could produce the same power at one-twentieth of the weight, with no worry about boilers, and this is the true measure of history.

47 SS *Great Eastern* (1858)—Paddle engine intermediate shaft
48 SS *Great Eastern* (1858)—Paddle engine room

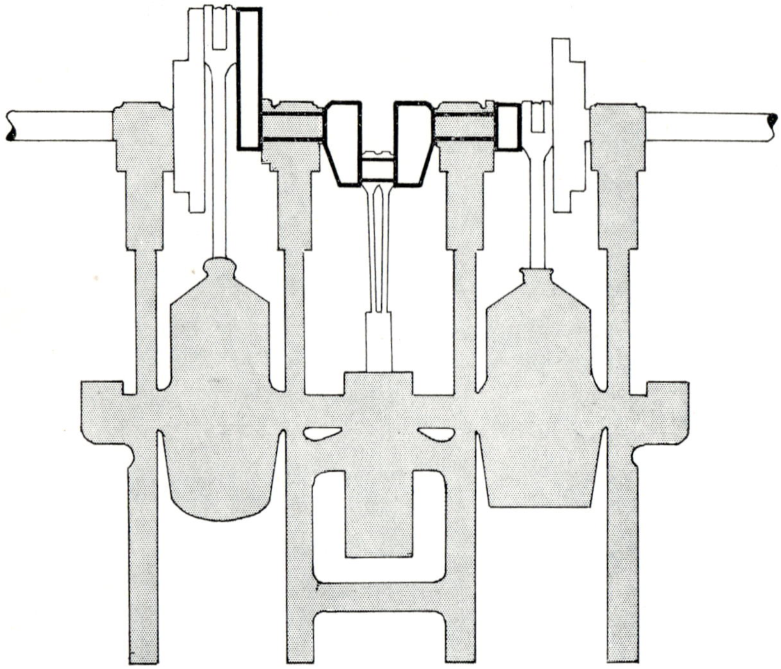

49 SS *Great Eastern* (1858)—Profile of paddle engine, showing intermediate shaft (*outlined*)

Finally, an unusual but most practical feature of this paddle engine was the toothed rack cut into the paddle-wheel circumference and which, when engaged by a loose hand pinion, served as a turning gear during overhaul. This was a refinement on the normal turning gear for paddle engines, which consisted of a pair of chain blocks hooked to the wheel spokes, one to pull, the other to check.

Whilst on the subject of paddle engines, just a quick look at three Dutch channel ferries built by the Fairfield Shipbuilding Co. in 1896. With memories of the Firth of Clyde passenger ferries and the English Channel pleasure steamers in the years between the wars, one recalls the elegant little diagonal paddle engines as pretty, polished models with a total horsepower measured in hundreds. The Dutch channel ferries *Koningin Wilhelmina*, *Koningin Regentes* and the *Prins Hendrik* were driven by gigantic diagonal engines that were anything but model size: triple-expansion engines with three cranks at 120°, the cylinder bores being 51 inches, 75 inches and 112 inches, by 78 inches stroke.

Steam was supplied at 170 lb and each set of engines developed 9000 IHP, giving a speed of 21 knots.

The real giants of the steam age, however, were not the paddle engines of the mid-century, nor yet the horizontal screw engines of a later decade: they were the tandem compound monsters in the 'seventies and 'eighties, when ship's speed and ship's size were being augmented at an ever-increasing rate without any compensating development in marine engineering techniques. To increase the power of an engine there was no provision for 'changing gear'—triple expansion had not yet arrived, the boiler pressure was too low to make it worth while.

Also there was no alternative drive available, no high-speed turbines, multiple diesels or, in the earlier years of the compound engines, no possibility of fitting twin screws. One tandem-compound engine filled the engine room completely, leaving no room at all for a second unit. The only way left to the shipbuilder was to fit a still bigger engine using the same boiler pressure as in the earlier ships. This led firstly to enormous-diameter LP cylinders, the largest, I believe, being 120 inches bore, and to lengthy piston stroke, then to the splitting of the LP cylinder into two separate cylinders of equal bore to keep sizes and weights within practical limits. And remember, in those days everything was handled by block and tackle.

The *City of New York* was re-engined in 1878 by just such a monster engine, a single-screw, four-cylinder tandem-compound engine on two cranks which just about filled the engine room like a foot inside a shoe, and extended well into the skylight casing. The two HP cylinders were 40 inches bore and the two LP cylinders 71 inches, the HP cylinder feet resting on columns high above the LP cylinders, and extending up from the main columns. Steam at 75 lb was supplied by four double-ended boilers in two stokeholds, each boiler space having its own funnel, and the rotational speed was a modest 52 rpm giving a purely nominal horsepower of roughly 2600, corresponding to, say, 6000 IHP by modern reckoning

All pumps were driven by beam levers from the two crossheads, with the exception of a couple of banjo general-service pumps at the forward end, and the condenser was placed below the main engine bedplate, surely a most unusual position in a reciprocating engine. A most noteworthy feature in the engine design was the provision made to draw the LP piston without disturbing the HP monkey cylinder, which was effected in the following way. The HP and the LP pistons were connected by a single piston rod passing through the LP top cover and the HP bottom cover, the rod being coned into both pistons from below and secured by nuts on the upper faces in the usual way. The LP piston drove the

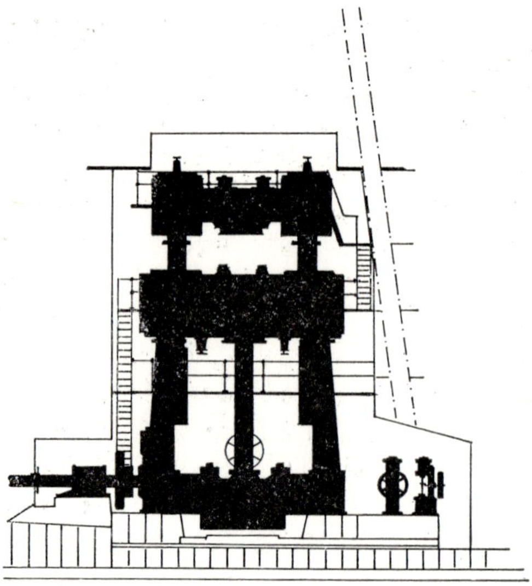

50 SS *City of New York* (1878)—Section through engine room

crosshead by two piston rods through stuffing boxes in the bottom cylinder cover. There was a manhole in the centre of this bottom cover, and another in the top cover, set to one side.

To remove the LP piston, the HP top cover was removed together with both manhole lids in the LP covers. The nuts on the single piston rod were then removed and the rod dropped down through the LP cylinder into the crankpit. The nuts on the two LP piston rods were then taken off and with the LP top cover taken to one side, the LP piston could be drawn out and pulled clear between the two upper frames. This operation was an important improvement in vertical tandem engines, the removal of the LP piston for routine examination usually being a dockyard job requiring the monkey cylinder to be lifted clear of the engine and placed temporarily on a reinforced platform.

The engine-room complement for this huge engine was six engineers, there being twelve firemen and six trimmers, and it is wondered how the firemen divided the work among themselves. Four double-ended boilers are considered to be the equivalent of eight single boilers, and manning practice requires one fireman per boiler per watch, making a total firing personnel of twenty-four. Evidently in the 1870's ships' firemen had to work very much harder than they did in my days at sea.

The conventional triple-expansion engine as we know it, i.e. with three cylinders

51 SS *Teutonic* (1889)—Middle platform of starboard engine

on three cranks at 120° to each other, reached their apogee in the last remaining years of the nineteenth century. After that period, bigger or more powerful engines were obliged to have the LP stage carried out in two cylinders, or separately in an exhaust turbine. The two Atlantic giants *Majestic* and *Teutonic*, the first of the White Star twin-screw ships, had a pair of three-cylinder triple-expansion engines with cylinders HP=43 inches, IP=68 inches, and LP=110 inches, the common stroke being 60 inches. Each engine developed 8500 IHP, giving a ship speed of $19\frac{1}{2}$ knots, there being twelve double-ended and four single-ended boilers to provide steam at 180 lb.

These two ships were launched in 1889, and by 1908 they, together with the American *New York*, were still the largest triple-expansion three-cylinder engines yet built.

In 1893 Cunard brought out two record-breakers on their Atlantic ferry service: the *Campania* and *Lucania*, built by Fairfields on the Clyde. They were twin screw, developing a total of 31,000 IHP per ship, the engines rotating at 84 rpm. Each engine worked in its own engine room, there being a centre-line bulkhead in the engine room extending to the main deck, and consisted of five cylinders on

52 SS *Lucania* (1893)—Upper platform showing cylinder tops of both engines

three cranks, the two HP cylinders being tandem to the LP cylinders, with the IP cylinder between the two LPs.

Dimensions for each engine were: two HP at 37 inches, one IP at 79 inches, two LP at 98 inches, the common stroke being 69 inches. The height from bottom of bedplate to top of HP cover was 47 feet, and the HP cylinders each had their individual grating and access ladder extending into the skylight casing.

The crankshaft, 26 inches in diameter, weighed 110 tons, and each thrust shaft, 14 feet long, carried fourteen collars.

The engine pumps were duplicated in each engine room, there being a set of beam pumps driven off each LP crosshead, and the main condenser was divided and could accommodate each LP cylinder exhaust independently.

The cylinders were steam-jacketed in accordance with the usual practice of the day and manœuvring was by Brown's steam-hydraulic gear. The governor gear controlled the operating valve of the reversing gear instead of the throttle valve, in an emergency the links would be run over into neutral and the engine would stop in a very few revolutions on compression alone, although in so stopping, the LP cylinder relief valves would probably lift and stay lifted. With

53 SS *Caledonia* (1894)—Engine room top platform

normal governor control by throttle, the steam would be shut off the engine but the way of the ship would keep the propeller turning and it would take many minutes before the engine came to rest. In a twin-screw ship the sound engine, if left running, would drive its defective twin engine indefinitely through the medium of the propeller in the slip-stream.

The *Lucania* and *Campania* had twelve double-ended and two single boilers, using steam at 165 lb, and they were the first vessels on the Atlantic run to use lap-welded wrought-iron steam piping instead of copper. These pipes were 20 inches bore, and the manœuvring valve for each engine was to the same gigantic scale. A separate steam engine was used to open and close these valves, and it was possible for an agile man to crawl through the valve chest.

An engine not unlike those just described was fitted in the P. & O. steamer *Caledonia*, built in 1894. This was a single-screw engine of 11,000 IHP with three cranks, there being two HP cylinders 33 inches bore, one IP at 69 inches and two LP cylinders of 84 inches each, the common stroke being six feet. Piston rods and crosshead pins were both $11\frac{1}{2}$ inches diameter, and the engine thrust was taken up by two thrust shafts in tandem, each with thirteen collars. The HP piston valves were worked through rocking levers by the LP slide valve and this served a dual purpose. It transferred the LP valve movement towards the HP valve centre line and it helped to counter-balance the dead-weight of the LP slide valve, $1\frac{1}{2}$ tons in this case.

A quaint little sidelight on this large luxury steamer: her electrical installation consisted of three compound engines driving dynamos with a total of 69 kW.

The boiler installation seems reasonably modest by previous standards, there being three double-ended and four single-ended boilers in all.

So far, the huge engines described above were installed in British ships which in the main served on the Western Ocean run, and it is true that the giants and record-breakers were confined to the luxury trade between UK and USA. But not all these ships were British built or run, and some of the German steamers at the turn of the century equalled, and in some cases surpassed, their rivals both in speed and power, and sometimes in luxury.

The North German Lloyd steamer *Kaiser Wilhelm der Grosse*, built in 1897 by the Vulcan Co. of Szczecin, was one of the first 'four-stackers' on the North Atlantic ferry service, and came of a long line of emigrant ships dating back to the original *Bremen* of 1857. In these days of high-pressure high-capacity boilers, even the largest ships can get by with two or three water-tube boilers only, their uptakes all combining in one funnel. But at the turn of the century, when Scotch boilers were the only type available in passenger ships, the large powers then in demand required a very considerable number of boilers placed in separate stokeholds, each group having its own funnel. Another aspect of funnel capacity lay in its impact on the Eastern European peasant mind: the *Kaiser Wilhelm der Grosse* carried 800 emigrants, and four funnels were a guarantee not only of immense power but of immunity to drowning.

Now to the marine engineer of my generation four funnels implied quadruple screws, and although there has never been a ruling on this matter, generally speaking the four-funnelled ships of the past have usually been driven by an equal number of propellers, the *Olympic* and *Titanic* notwithstanding. However, in spite of this dictum, the *Kaiser Wilhelm der Grosse* was twin screw only, the engines being four-cylindered triple-expansion units balanced on the Yarrow-Schlick-Tweedy system. Contrary to British practice, whereby the two LP cylinders are placed at the outer ends, the present engines had their cylinders placed in the normal flow path: HP, IP, LP and LP, one of the LP cylinders being made extra heavy to aid the balancing. Cylinder sizes were HP = 52 inch, IP = $89\frac{3}{4}$ inch and two LP = $96\frac{1}{2}$ inch, the stroke being $68\frac{7}{8}$ inch, and in keeping with the practice of the time, each piston had a tailrod projecting through the upper cylinder cover. Steam at 178 lb was supplied by twelve double-ended and two

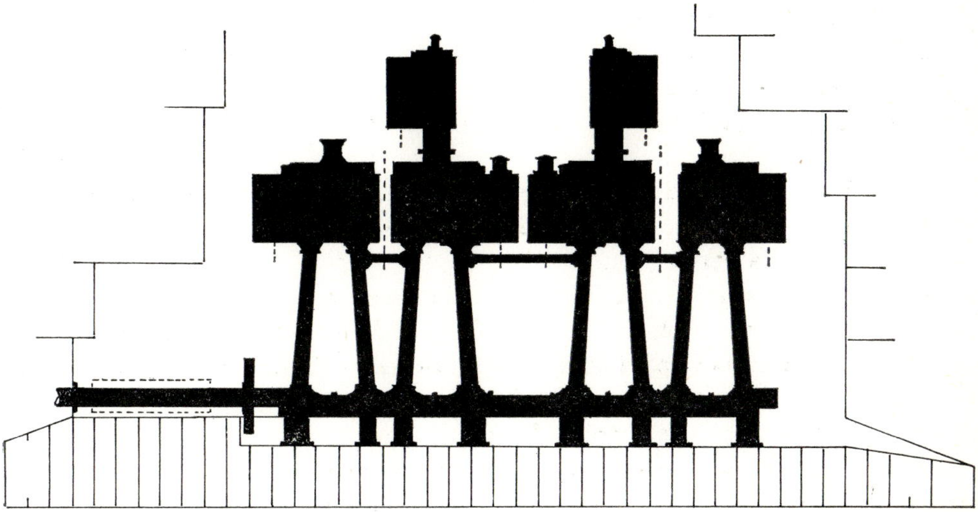

54 SS *Deutschland* (1900)—Port main engine

single Scotch boilers, in four separate stokeholds, each boiler room having a pair of electrically driven forced-draught fans.

With 27,000 IHP, this vessel was the largest and fastest of her day, and owing to her fine lines and quick run aft, the propeller discs overlapped by about 8 inches. The port propeller was therefore placed some 3 feet forward of the starboard to clear each other in the stern-frame aperture.

The total height of each engine was 40 feet and each enormous crankshaft weighed 82 tons machined. All engine-room pumps were independent with the exception of a sanitary and bilge pump driven by a pin at the forward end of the shafting. The independent air pumps were built on the same monumental scale. They were duplex type with twin steam cylinders, there being a horizontal shuttle valve between them worked by the beam. The total height of these pumps, not including the seating, was 13 feet 7 inches, and they required their own access ladders and gratings.

Another, similar vessel to the above was the twin-screw steamer *Deutschland* built in 1900 by the same shipyard for the Hamburg America Line. This vessel also had four funnels, placed as two separate pairs in the German fashion instead of the more usual four equally spaced, to accommodate her twelve double-ended and four single-ended boilers. Her two screws were driven by six-cylinder quadruple-expansion engines on four cranks there being two HP cylinders in tandem with two LP cylinders, with the first IP forward and the second IP at the after end.

The cylinder sequence was thus, from forward to aft: first IP, HP+LP, HP+LP, second IP. Cylinder diameters were two HP at 36·6 inches, first IP at 73·6 inches, second IP at 103·9 inches, two LP at 106·3 inches, with a stroke of 72·8 inches, and each engine developed 16,500 IHP at full power, with a steam pressure of 220 lb.

As usual with these giant engines, there were no engine-driven service pumps other than a small bilge and a small sanitary pump worked from a pin at the forward end of the crankshaft.

All cylinders other than the two LPs had piston valves, while the LP cylinders had two flat slide valves each, and one of the odd features of this engine was the provision of six sets of valve gear for a four-crank engine. The normal arrangement with monkey cylinders was to have their piston valve operated by the same eccentric rods and spindle as worked the valve for the lower cylinder, movement from lower to upper spindle being translated through a rocking beam on top of the lower cylinder. In the *Deutschland*, however, the two HP monkey valves were operated by their own eccentric pulleys, rods and spindles.

The crankshafts were of 25 inches mean diameter with a 10-inch central hole, and, again following German practice, there was a muff coupling between the intermediate shafting and each tail shaft, to allow of the latter being drawn out into the drydock for examination and so save disturbing the line of shafting by pulling it inboard. In passing, this huge coupling whirling round in way of the stern gland required special guard plates and railings, and was generally most difficult to release. In practice the screwshaft was drawn inboard in the usual old-fashioned way with the coupling and adjacent shafting still attached.

The *Deutschland*'s engines were separated by a longitudinal bulkhead which, in effect, formed two engine rooms, and they were controlled by servo-motor-operated manœuvring valves, similar to those fitted in the *Campania* and *Lucania*. The Aspinall governor was arranged to act on the hunting gear of this servo motor rather than on the main valve itself.

Finally to complete this trio of German giants, we have the North German Lloyd steamer *Kaiser Wilhelm II*, again a four-funnelled ship driven by twin screws and again built by Vulcan Co. of Szczecin in 1902. Her engines were of a type that defies nomenclature, but basically each engine consisted of an eight-cylindered quadruple-expansion engine driving six cranks, the HP cylinders working in tandem on the first IP cylinders.

Each propeller was driven by a pair of three-crank quadruple-expansion engines connected together on the same crankshaft, and the total power developed on each shaft was 20,000 IHP, the maximum ever achieved by a marine steam

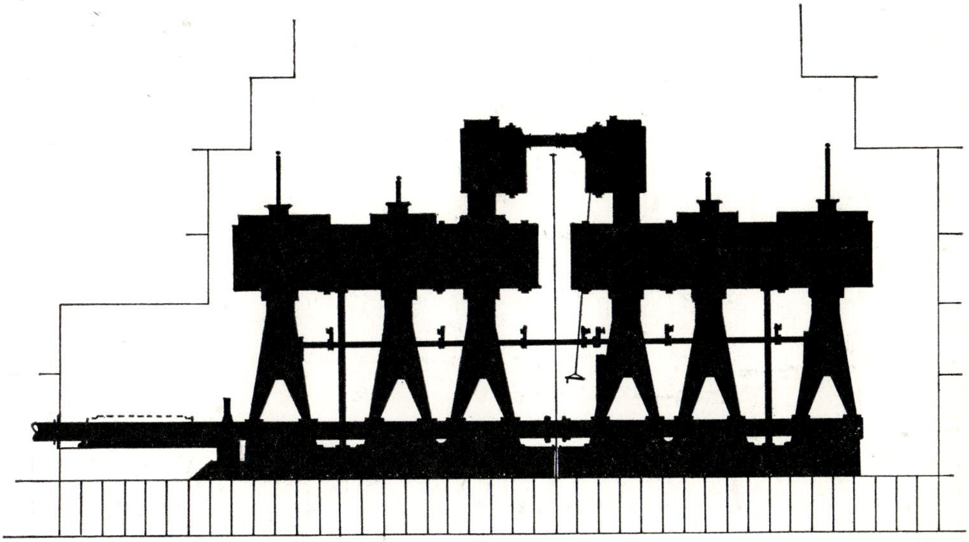

55 SS *Kaiser Wilhelm II* (1902)—Port main engine

reciprocating engine. Although a purist might consider this power to be supplied by two separate engines working in tandem on the same shaft, the fact remains that the combined engine was controlled and manœuvred as one. The ship's engine room was divided both longitudinally and athwartships by a centre-line bulkhead and a transverse bulkhead between each half-engine. In the latter case both the main shaft and the reversing weighshaft, which were continuous for the two half-engines, pierced the bulkhead through watertight stuffing boxes.

Although these two double engines normally ran as twin engines, in the event of damage to the forward section it could be disconnected and the after section run on its own, and each of the four component engines had its own condenser, air and feed pumps.

An interesting point to note is that the A frames, of cast steel, were not considered to be sufficient support for the overhang of the LP cylinders and valve chests, and forged steel pillars were fitted to supplement the framework.

Cylinder bore and stroke data were as follows: HP = 37·4 inches, first IP = 49·2 inches, second IP = 74·8 inches, LP = 112·2 inches, stroke = 70·8 inches, the crankshaft being 25 inches in diameter and 72 feet long, turning at 80 rpm. Steam at 213 lb was provided by twelve double-ended and seven single boilers in four stokeholds and the total horsepower developed by the two double engines was 40,000.

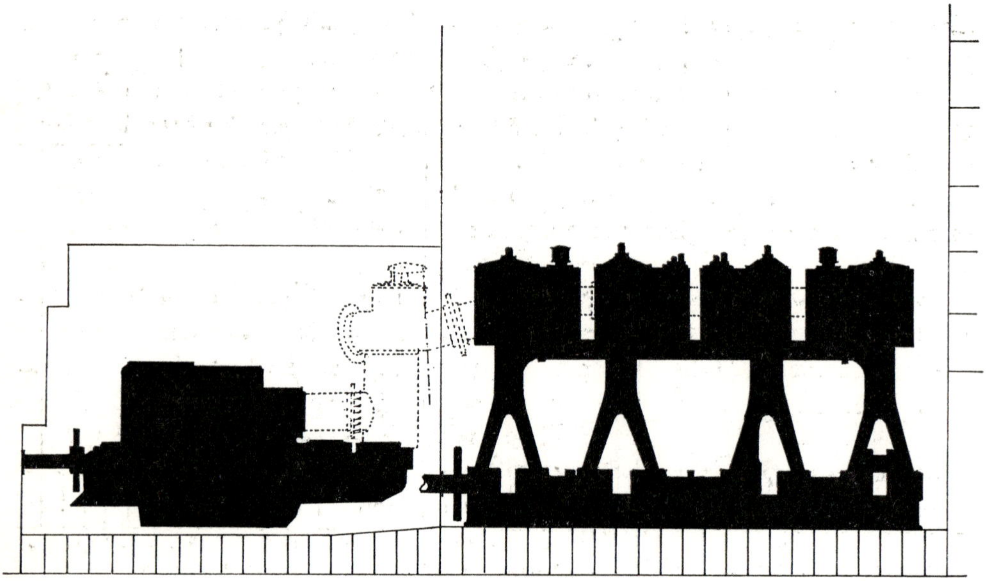

56 SS *Olympic* (1910)—Port wing engine and centre exhaust turbine

Back on this side of the channel and a few years later the White Star Line ordered three giants from Harland & Wolff Ltd in Belfast: the four-funnel triple-screw *Olympic*, *Titanic* and *Britannic* were designed to be the largest and most luxurious vessels on the North Atlantic and they were driven by the largest marine steam reciprocating engines and exhaust turbines ever built. The *Olympic*, built 1910, set the pace and the ill-starred *Titanic* followed in 1911, with the *Britannic* bringing up the rear. All were propelled by similar engines: a pair of triple-expansion four-crank wing engines both exhausting into a direct-coupled low-pressure turbine on the centre shaft, the wing shafts running at 77 rpm, the centre one at 170 rpm, and a total of some 50,000 IHP was developed by the three shafts. Wing engine cylinders were as follows: HP at 54 inches bore, IP at 84 inches and two LP at 97 inches, the common stroke being 75 inches, with piston valves for all cylinders.

These immense engines exhausted through change-over valves, under overriding control by the Brown's reversing gear, either into the central turbine or direct to the condenser, the change-over valves themselves being almost the size of the LP cylinders. Thus the exhaust turbine could either drive forward or it could idle, but it was not constructed to run astern. The main stop valves on the reciprocating engines were 21½ inches bore and were far too large to be conven-

57 SS *Olympic* (1910)—Centre exhaust turbine rotor

iently operated by hand during manœuvring; a smaller, auxiliary valve was therefore introduced, bypassing the main valve, which was used whilst handling the engines. The main valves were opened only when 'Full Away' was rung down.

Some of the *Olympic*'s engine vital statistics are shown and will give an idea of the problems involved merely in carrying out routine overhaul. As in most four-crank triple-expansion engines with the LP steam split into forward and after cylinders of identical size, the scantlings of the running gear for the outer cylinders was made somewhat lighter than that for the two inner engines, the HP and the IP. But for the latter engines, the dimensions were indeed gigantic:

Diameter HP and IP piston rods	14 inches
Diameter HP and LP connecting rods	13½ to 15 inches
Diameter HP and LP crossheads	16¼ inches
Diameter crankpins	27¾ inches
Diameter crankshaft journals	27 inches
Thickness crankwebs	20 inches
Length crankpin between webs	35 inches
Diameter crankpin bearing bolts	8½ inches

The complexity of the problems may now be appreciated: the crankpin bearing, $27\frac{3}{4}$ inches in diameter and nearly a yard long, flailing round on a 6-foot 3-inch crank circle at 77 rpm and transmitting alternate thrust and pull, centrifugal and inertial forces of many, many tons, had to be adjusted to the thickness of a piece of paper. This adjustment was effected by a couple of nuts so heavy that each could not be lifted by less than three men, and the tightening was done by flogging them up on a swinging tup weighing several hundredweights.

The exhaust turbine was no less impressive in size, and presented a mystery to the non-technical mind by using steam at 10 lb absolute pressure, i.e. already a healthy vacuum to most people, and exhausting it at 1 lb absolute. The overall length of the rotor was 50 feet, the diameter being 17 feet. Weighing 150 tons, it ran at 170 rpm and developed about 18,000 SHP.

Another pointer to the *Olympic*'s enormous size lay in the use of her main injection-valve openings in the ship's side as a workmen's access doorway when she lay on the slipway before launching.

Steam at 215 lb was supplied by twenty-four double- and five single-ended boilers, and final proof of her size was given by marine engineers on the Western Ocean who swore that it took at least one of those boilers to provide the steam that leaked past the HP glands.

There were also giants among the turbines in those days of direct drive, but they were not as spectacular as the reciprocators. Even to the professional eye a turbine running at full speed looks just the same as a stationary one—there is no movement to catch the eye. To an engineer of my generation the sight of a well-maintained steam reciprocating engine running at full speed was indeed poetry in motion, and as different from a turbine installation as a coloured motion picture is from a lantern slide.

Two of the most famous of the old direct-coupled turbine ships were the twins *Mauretania* and *Lusitania*, the former built by Swan Hunter and Wigham Richardson Ltd, at Wallsend, and the other by John Brown & Co. Ltd, of Clydebank, both in 1906. They were quadruple-screw vessels, with hollow bored shafting running at 190–200 rpm and developing 70,000 SHP total output on trials, giving a top speed of 26 knots.

As built, all four shafts had three-bladed propellers, but later, to correct slight vibration, the wing propellers were removed and replaced by four-bladed ones.

The engine-room arrangement was as follows: HP turbines on port and starboard wing shafts, with LP and astern turbines on the two inner shafts. Steam

58 SS *Mauretania* (1906)—LP turbine casing in foreground, next to the astern casing, HP turbine casing in middle distance and thrust carriage in background

at 180 lb was supplied at 200 lb by twenty-three double-ended and two single-ended boilers in four stokeholds, there being a separate funnel to each group of boilers.

An idea of the immense size of the turbines is shown in the accompanying illustration by comparing their casings with the workmen standing alongside them. The LP casing is shown on the foreground next the astern casing and in the middle distance the HP casing, with the thrust carriage beyond. The HP casing was 25 feet long by 10 feet internal diameter, while the combined LP and astern casings stretched for 90 feet, the LP rotor being 16 feet 6 inches in diameter, and the rotating mass weighing 200 tons.

Some interesting statistics quoted by engineering magazines of the day point to the power and consequential cost of propelling a ship of 38,000 gross tons through the water at 26 knots. At full speed 375 tons of steam per hour passed through the machinery, and to condense this steam 450,000 tons of sea water per day had to be pumped through the condensers, assuming a circulating/feed water ratio of 50. This represented the water supply for a city the size of Paris, with a population of 3,000,000 in 1907. With a coal consumption of 1000 tons per day, the air required to burn the coal in the furnaces was 20,000 tons per day, or 26,100,000 cubic feet per hour.

The old *Mauretania*, probably the best-known ship ever to sail the Atlantic, was a happy vindication of Froude's law of mechanical similitude. In 1890, when the *City of Paris* and the *City of Rome* represented the last word in cross-Atlantic travel, being 560 feet long with 13,000 tons displacement and developing 18,000 HP for a speed of 20 knots, Froude pointed out that a vessel of 800 feet at 40,000 tons displacement and speeding at 25 knots would need engines of 65,000 HP. These are roughly the characteristics of the *Mauretania* and the *Lusitania*.

Giant marine engines were also built on the other side of the Atlantic, not only for ocean-going ships but also for the indigenous river craft dealt with in a previous chapter, the Eastern and the Western River steamboats.

One of the largest was fitted into the *Joseph B. Williams*, a Mississippi stern-wheeler, built in 1876, and consisted of a pair of horizontal tandem-compound engines, one on each side of the deck, coupled to the stern-wheel barrel shaft. Each HP cylinder was 19 inches bore, the LP 44 inches with a common stroke of 9 feet, and the HP and LP cylinders formed a common casting, bolted direct to the wooden base. The HP piston had one piston rod, the LP piston two, and all three rods coupled to the one wide crosshead, from which a 53-foot long wooden connecting rod transmitted the motion to the stern-wheel cranks. This stern wheel was 29 feet in diameter and 29 feet wide, with paddles 38 inches deep. The wheel crankshaft was 35 feet long by 15 inches in diameter, and weighed 25 tons.

Six highly rated locomotive-type boilers supplied steam at 140 lb, the boilers being placed in a group fore and aft at the forward end, mounted on brickwork, and cross-connected from below by a pair of athwartships mud-drums. These boilers thus formed a single steaming unit.

Contrary to the usual practice, this engine had no coupled pumps, they were worked separately by an auxiliary beam engine situated alongside the boilers.

But the real American heavy-weights were to be found in the fast passenger ferries on the New York–Albany run, engines that in their own way could compare with any on the Western Ocean throughout the nineteenth century.

The *New World*, built away back in 1849, was propelled by just such a monster, a side-paddle engine of the timber-gallows frame type. The single cylinder was 76 inches in bore, the piston stroke being 15 feet, the greatest piston stroke ever recorded in any marine engine. The wooden paddle-wheels were 45 feet in diameter with floats 11 feet 9 inches long by 3 feet deep. The lozenge-shaped walking beam was 26 feet over centres and 13 feet deep in the middle. The piston rod was

59 American paddle steamer *Puritan* (1889)—Forged strap for engine walking beam

8 inches in diameter while the main connecting rod was 10 inches thick, reinforced by 2-inch truss rods stretched over struts along its length, and 28 feet long.

This vessel literally stabbed herself to death when the connecting rod neck snapped at the connection to the walking beam. The rod smashed its way through successive decks and ploughed through the bottom of the ship.

Perhaps the biggest single-cylinder engine ever built was fitted into the *Pilgrim*, launched in 1882 for the Albany run. This engine was also of the gallows-frame type, and developed 6500 IHP on 50 lb of steam in a cylinder 110 inches in diameter with a stroke of 14 feet. The working beam measured 29 feet between centres and weighed 33 tons, while the 13-inch-diameter connecting rod weighed 12 tons.

The crankshaft, in two sections of 40 feet each, weighed 80 tons and was laboriously built up of old scrap metal, horseshoes, bolts, boiler plates, etc., into a 30-inch billet in the fashion of the time. This type of 'forging up' into a shaft 30 inches diameter made for a very vulnerable forging; an earlier vessel of similar proportions, the *Providence*, broke her crankshaft twice in the same year.

In this particular engine the gallows trestle was built up of iron box section and not of the more usual timber.

Steam was provided by twelve boilers arranged in four batteries of three each, and the ship carried an engine-room crew of thirty.

Probably the largest and most powerful walking-beam engine ever was that driving the side-wheeler *Puritan*, built in 1889 for the Fall River Line and in service on the Long Island Sound run. The ship was built by the Chester Ship & Engine Building Co. and engines by W. & A. Fletcher & Co.

Contrary to the usual run of walking-beam engines, this engine worked on compound expansion, taking steam at 100 lb from eight boilers and it developed 7500 IHP at full speed of 21 knots. The two steam cylinders were positioned along the centre line of the ship, the LP directly below the end of the walking beam and the HP nearer to the centre. The HP cylinder was 75 inches bore with a stroke of 9 feet, the LP was 110 inches bore with a stroke of 14 feet. The piston crossheads for the two cylinders connected by separate double links to the walking beam, which consisted of the usual cast-iron skeleton with a wrought-iron frame.

The paddle-wheels, 35 feet in diameter and 8 feet wide, were of the feathering type, and revolved at 20–24 rpm. Each wheel weighed 100 tons.

Some idea of the size of this engine is given by the following list of dimensions and weights.

Total height of engine from base	71 feet (i.e. the height of an eight-storey building)
Walking beam—centres	34 feet
weight	43 tons
trunnion diameter	18 inches
Crankshaft main bearings diameter	30 inches
Connecting rod length	39 feet
Eccentric throw: HP	18 inches
LP	24 inches

All these giant steamers, or nearly all of them, were on the Western Ocean run, the most exacting service in the world of shipping. The North Atlantic is not a geographical area so much as a meteorological hotchpotch. It seems to get more weather than anywhere else, and all of it bad. If the westerly gales abate for a day or two, the fog clamps down or the ice-floes sweep in from the north. To make matters worse, this ocean is one of the busiest in the world, with large and luxurious liners among its commuters, all of them running to a very tight and rigid schedule—cargo and mail must go through and must arrive on time.

This mailboat service bred an extremely tough and efficient type of sailor, both deck and engine room, who was prepared to risk his comfort, his certificate, career, even life itself, to ensure his ship catching the tide. On deck, in spite of head seas, fog or ice, the Master just had to bash his way through to keep his schedule and his job, with only a token reduction in speed.

Down below a similar urgency prevailed once the pilot was dropped: the engines had to be kept running whatever else happened. There was nothing, just nothing, in the whole gamut of engine defects that could excuse slowing down for adjustment, and such minor worries as choked bilges, smoking bearings, loss of vacuum or leaky steam-pipe joints just had to be put right before the watch could be handed over.

Then there was the matter of coal. These giants burnt up to 100 tons of the stuff every watch, coal that had to be dug out of a 'tween-deck bunker away up forward by a crew of troglodytes who never saw the light of day, laboriously wheeled to a small hatch against the roll of the ship and tipped down into a lower bunker from where it was shovelled by another gang on to the firing platform. From this point it was processed in the furnaces in a strictly regimented and unceasing rhythm to the beat of a gong: 'Fill her up boys'—gong—'Go ahead on the rake'—gong—'Two up off the bars' (with the slice)—gong—'Rake it back'—gong—etc., etc. Again, burning 100 tons of coal a watch left many tons of ashes that had to be discharged overboard through an ejector. As often as not, this ejector choked and spilled the ashes on to the tank-top; the ensuing choked bilges promised a lively half-hour of bilge-diving to clear them after the ashes had been cleared away.

Sometimes an 'old man', a fused mass of clinker the size of a sheep, would form in the back end and would have to be broken up before it could be drawn out through the fire door. Breaking it up involved hard work with long rake and slice in the searing heat of the furnace mouth, and on occasion this wrestling with large lumps of red-hot clinker dislodged the firebars into the ashpit; then the officer of the watch was in real trouble and could anticipate another hour of hard sweaty work getting the furnace going again.

This type of dedicated seaman, the Western Ocean sailor, may not have been appreciated by his seniors or his company, but informed opinion ashore seemed to consider him much superior to his opposite number in the Navy. Here is a modest tribute paid him by a member of Parliament and quoted in an editorial written towards the turn of the nineteenth century:

'If twenty first-class ships of our mercantile marine were despatched tomorrow from England to New York, with orders to proceed thither at the highest speed of the slowest ship, say 17 knots, the squadron to keep together as much as possible, it is all but certain that these twenty steamships would cross the Atlantic without a hitch or a breakdown in their machinery, the weather being neither worse nor better than it usually is at the time of the year. If twenty of our best warships, including cruisers, were ordered to make a similar crossing at the highest speed of the slowest ship, say 15 knots, it is about equally certain that more or less serious casualties would occur, and that some of the ships would either arrive on the American coast in tow of their consorts, or else finish their voyage at about half speed. The truth of the first proposition is proved every day by the ordinary trips of the vessels of the Cunard, White Star, International and other lines. The truth of the latter proposition is demonstrated by the record of the naval manœuvres.'

10 Diesel engines

Although most of the public and much of the Press consider Dr Diesel to be the inventor of the oil engine, on the assumption that the heavy-oil engine in universal use on land and sea is named after him, it would not be fair to the many inventors associated with the development of the internal combustion engine to subscribe to this view: people like Lenoir, Otto and Langen, Beau de Rochas, Daimler, Akroyd-Stuart, Clerk and many others in a long chain of inventors, starting with William Barnett, the shadowy figure behind the original patent for the gas engine in 1838. Rudolf Diesel's part in the history of the oil engine was very important indeed as his researches into compression ignition enabled him to run an oil engine without having recourse to a troublesome magneto or a dangerous blowlamp to fire the cylinder charge. But it was still only a part; oil-vapour engines were in general use when Diesel was a little boy.

However, whether we agree or disagree with public opinion, the fact remains that the ubiquitous compression-ignition engine will continue to be called the diesel engine, and will be referred to as such in this chapter.

Although petroleum, under a variety of names, was known to the Ancients, its use as a fuel was never seriously considered mainly because it was seldom found in sufficient quantities to warrant collecting, and, when found, it usually appeared as a non-inflammable black ooze. However, the petroleum industry as we know it started away back in the 'fifties of last century by the sinking of the first tube-well at Titusville, Pennsylvania, in 1859 and at Ploiesti in Romania in 1857. Much of this product, the 'rock-oil', was used for illumination and cleaning, but it was soon found that its distillates, being highly volatile, could also be converted into an air-vapour mixture and used in a gas engine.

The gas engine of the 1860's worked on the Lenoir cycle, wherein the air-gas mixture was drawn into the cylinder on the suction stroke for about one-third travel, when the inlet valve closed and the mixture was exploded by contact with an open flame, the remaining two-thirds of the stroke forming the power stroke. The piston return made the exhaust stroke. This cycle gave very little power at a very slow speed, about 200 rpm, and could not seriously compete with the steam cycle, especially as the engine depended on a source of gas, either coal or producer, for its fuel. It was, inevitably, a hazard inasmuch as ignition of the gas charge depended upon a naked flame housed in a recess in the cylinder head, actual firing being made through a small flap valve opening inwards into the cylinder by the suction.

The arrival of a petroleum distillate on the market, in 1873, enabled it to be used in a gas engine, suitably vaporised by drawing air over a surface of petrol, and the gas engine thus became a self-contained oil-vapour engine that could be transported anywhere. The original method of vaporising consisted of a wad of teased asbestos lodged in a recess in the air inlet valve pocket, kept saturated by a slow feed of petrol, and protected from the naked flame of the pilot jet by a gauze wire screen. In time, this petrol-soaked wad developed into the modern carburettor after many metamorphoses, but the principle remains the same; a current of air drawn over a surface of liquid or atomised petrol.

The petrol motor led a somewhat haphazard existence for a number of years, not good enough to use but too good to throw out, until Daimler brought out his revolutionary petrol engine in 1883, and which formed the prototype for the motor-car industry. Instead of trundling round at 200 rpm, with just sufficient power to turn the flywheel over centres, Daimler's engine ran at 800 revolutions on the positive Otto or four-stroke cycle, and to ensure adequate lubrication, all the running gear was enclosed in an oil-tight crankcase. He fitted a float carburettor and regulated the engine speed by a 'hit and miss' governor controlling the exhaust-valve closing period. Another unusual feature in his engine was the grouping of both inlet and exhaust valves in one chest, thus reducing unnecessary clearances.

However, one of his most important contributions was the use of an ignition tube placed in the air-inlet pocket to ignite the cylinder charge. This tube, closed at its outer end, protruded through the wall of the cylinder head and was heated from outside the engine by a blowlamp mounted under a hood. Another item of equally outstanding significance was the introduction of the ratchet starting handle. Hitherto, the gas engine and its cousin the petrol-vapour engine had to be started by jerking the flywheel round by hand or by stamping down on the spokes; not too dangerous an operation on a Lenoir engine with no compression, but one not to be recommended on a compression engine.

Paraffin oil, being a heavier distillate than petrol could not be vaporised in a cold carburettor and so far had not been used as a substitute for gas. However, about 1889 Kjelsberg brought out his paraffin engine where the fuel was warmed and aerated in an exhaust-heated vaporiser, firing being by ignition tube, which still required a separate flame to keep it hot. However, electric ignition, first introduced in 1883, soon became the acknowledged means of firing the cylinder charge in the petrol and paraffin engines, although the flame-heated ignition tube was retained for the heavier oils.

Another link in the long chain of oil-engine development was the hot-bulb engine brought out by Akroyd-Stuart in 1890, in effect an extension of the hot-tube ignition engine, though with a difference. This engine, of the low-compression type, had an uncooled combustion antechamber in the cylinder head, communicating with the cylinder by a narrow passage. No vaporiser was necessary, and the fuel was injected into this antechamber through a needle valve by a jerk pump. This combustion chamber was heated initially to a dull red heat with a blowlamp, which was removed as soon as the engine took up load, the heat of combustion keeping the surfaces hot enough to ensure steady ignition. In this engine the explosive mixture was not generated during the suction stroke but during compression. The oil was sprayed into the hot antechamber which was already filled with exhaust gases remaining from the previous stroke, and it thus formed a non-inflammable mixture of oil vapour and exhaust gas, while the air for combustion was drawn directly into the cylinder through the inlet valve. On compression, air was forced into the combustion chamber to mingle with the oil vapour and form an explosive mixture sparked off at the correct instant by contact with the internal surfaces of the hot bulb.

Dr Diesel's experimental engine was built in 1894 by Maschinenfabrik, Augsburg, builders of the MAN diesel engine of later years, and although his patent required the engine to run on pulverised coal, there is no record that it ever did, or indeed that any piston engine could run efficiently on such a gritty fuel. This engine dispensed with carburettor or vaporiser, the fuel oil being injected and simultaneously vaporised by a blast of high-pressure air through an atomiser in the valve nozzle and ignited by the heat of compression alone. In short, Dr Diesel introduced the most important feature into the oil-engine industry which hitherto had been sadly lacking: safety. His engine was a heat engine in every sense of the word, with a high compression pressure and temperature and a high exhaust temperature, but it could run on fuel with a reasonably high flash point and it required no naked flame or electric spark to keep it running.

However, engineering is like medicine in that cures for specific troubles very often have their side effects, some as troublesome as the original. In doing away with the fire hazards inherent in the blowlamp and the electric-spark ignition, Diesel had to build an exceptionally robust engine with a heavy frame, large-diameter crankshaft and running gear. But above all, the fuel injection demanded high-pressure air from 600 to 1000 lb to be provided by a two- or three-stage compressor driven by the engine itself, forming a parasite load of up to 10% of the nominal engine power. And not only a compressor but all its concomitant apparatus of water coolers, high-pressure air receivers in duplicate complete with a complicated piping and valve system.

It says a lot for Dr Diesel's engine, originally designed to work on blast injection of fuel, that although the main engine compressor gave endless trouble throughout its life and became known as the junior engineer's bogey, it was still remarkably efficient and economical. This type of fuel injection is now happily obsolete, but engines working on the air-injection system were being built well into the 1930's and some old-timers were still in service at least until 1958, probably later.

Before dealing with the history of the diesel engine proper, a form of para-diesel engine which existed for a time side by side with its prototype, merits description. This was the Trinkler engine manufactured by the Körting Brothers of Hanover, a firm famous for its oil engines towards the turn of the century, and which was associated with engines, both for airships in 1887 and submarines in the 1914–18 war. This engine housed a small auxiliary compressor in the cylinder head with a differential piston, the greater area being open to the cylinder pressure and the smaller area direct to the fuel-valve atomiser. Compressed air from the cylinder compression stroke was forced into this auxiliary compressor through a port and the differential piston, worked by the main cylinder compression, simultaneously shut the port and further compressed the charge in the small auxiliary cylinder sufficiently to atomise the oil spray. It may be recalled that a similar system using main engine-cylinder compression was in use in the 1930's to operate the fuel pumps on the 'solid injection' principle in converted blast-injection marine engines.

The earliest marine diesel engine on record was a very small two-cylinder unit of 20 BHP fitted as an experiment in a canal barge shortly after the turn of the century, but the first commercial motor vessel to be built was the triple-screw tanker *Vandal* built and engined by Nobel Brothers of St Petersburg (now Leningrad) in 1904 and, surprisingly for these pioneering days, she was propelled by diesel-electric transmission, her diesel engines being non-reversible. Her motors developed about 400 BHP and she was used mainly on the Volga and in the Caspian Sea. She was evidently successful, as she was followed in quick succession by similar tankers *Emanuel Nobel* and *Delo*, also for use in the same waters.

These ships were confined to inland waterways and coastal navigation in southern Russia and possibly their machinery was not subjected to the rigours of a North Atlantic passage or to a long haul round the world. The *Vulcanus*, on the other hand, was built in Amsterdam in 1910 as a deep-sea tanker, and was engined by a six-cylinder Werkspoor diesel engine of 500 BHP working on the four-stroke cycle. She was of conventional tanker build with a turret deck, and could make 8·4 knots at a displacement of 1900 tons. Her engine, built by Nederlandsche Fabriek van Werktuigen en Spoorweg Materieel (hence the shortened version Werkspoor), had cylinders 400 mm bore by 600 mm stroke and turned at 180 rpm.

As the *Vulcanus* was the first motorship intended for use in open waters and one of the first to be fitted with a direct-reversible engine, it might be of interest to examine the latter in detail and note how a number of technical problems were tackled and solved by early motor engineers. The most important problem of all was that of finding a simple and efficient method of reversing the shaft rotation without the intermediary of electrical drive, the only alternative for a large ship in those days. Sulzer Brothers had designed a reversing mechanism for their engines as long ago as 1905, with Atlas following in 1907, but the *Vulcanus*' engine was one of the earliest large engines to be so fitted.

Following precedent, the engine was of the open-crankpit type, as marine engineers of the day were suspicious of any reciprocating machinery whose bearings could not be seen or felt. The cylinders were mounted in two blocks of three, with open-ended, water-cooled liners. The pistons, of the crosshead type, were air-cooled by a fan driven from the flywheel, the supply being conveyed to each piston by telescopic pipes. The three-stage blast-air compressor was driven by beam levers off the main engine crossheads, the LP on No. 1 column, the IP and HP on No. 2, the inter-coolers being water-cooled. These beams also drove the engine service pumps: bilge and sea-water cooling. Lubrication was by hand feed.

Reversing was carried out using two separate camshafts, one for ahead running, the other for astern, each mounted on one arm of a stationary right-angled bellcrank frame, this frame riding freely on a layshaft driven at the same speed as the crankshaft. The camshafts were driven by spur gearing from this layshaft at a 2:1 ratio. The layshaft incorporated two small cranks at right angles to each other, and was rotated by the main crankshaft through two long tubular eccentric rods, worked through a pair of eccentric sheaves whose throw equalled that of the layshaft cranks. This form of drive appears over-complicated for an effect that could be simply produced by skewgearing, roller chain or spur-wheel train, but it persisted with some makes of Werkspoor engine for as long as blast injection of fuel remained in fashion, with the minor difference that in later models, four driving rods were used, with cranks at both ends. To reverse the engine, it was sufficient to rotate the bellcrank frame round 90° to bring the reverse camshaft into running position.

The fuel-control system in this engine was integrated with the blast-air supply in an automatic regulating float tank, again a piece of equipment retained in later engines for the duration of blast injection generally. Fuel from the main engine-driven fuel pump was pumped into a floating accumulator tank supported on levers and balanced by a counterweight, the levers being connected to the fuel-pump suction valve in such a way that when the float vessel was about half full, it tipped about its fulcrum and shut off the fuel supply. Blast injection air was admitted to this float tank at the top to equalise the pressure throughout the injection system, and fuel flowed from the bottom of the accumulator to each injection valve through (a) governor control valve, (b) manual control valve and finally to (c) individual regulating/shut-off valves to each cylinder.

This fuel distribution supply, taken from the bottom of the float chamber, was protected from immediate failure by an arrangement of the pipe suction which corresponded in principle to the emergency petrol tank in a car. The fuel distribution pipe was connected to an internal standpipe of reasonable height, this internal pipe having a hole near the bottom. If, through failure of the fuel pump, the oil level dropped below the top of this standpipe, there would still be sufficient oil flowing through the hole in the bottom to maintain the engine running at reduced speed for some minutes and allow the engineer to start up the stand-by pump.

The engine room of the *Vulcanus* was rather spartan by modern standards. On the port side, a two-cylinder diesel engine of some 40–50 HP drove a two-stage compressor to supply air for starting and reversing, also to drive the four steam-pattern winches on deck, worked by compressed air through a reducing valve. This engine also worked the cargo pump through a clutch. Another auxiliary oil engine, of 10 HP, drove the single dynamo.

On the starboard side were four starting-air receivers charged at 300 lb pressure.

The *Vulcanus*, even in those early days, showed a remarkably good fuel consumption figure of 0·36 lb/IHP/hour, although she also experienced a number of accidents which recur time and again in the internal combustion engine. In 1912 a main-engine bottom-end bearing broke, the bearing keep fell into the crankpit and jammed the crankshaft, a new shaft having to be fitted at Barrow-in-Furness. Later, an air receiver burst and upon enquiry it was found to have been working almost full of water. These early air receivers were not fitted with relief valves, though after the accident bursting discs with spring-loaded safety valves were fitted all round.

The *Toiler*, built in Newcastle in 1911 with a Swedish diesel engine, was, I believe, the first cargo motorship to cross the Atlantic, although she was still of quite modest dimensions. She was unusual in that her steering gear was worked by compressed air, and her cabins were heated by an exhaust-gas boiler.

Another motor ship, Dutch built, was placed in service in 1912. This was the tanker *Juno* of 4300 tons displacement driven at $10\frac{1}{2}$ knots by a 1460 IHP oil engine running at 125 rpm. The six cylinders were 22 inches bore by $39\frac{3}{8}$ inches stroke and were mounted on a complicated system of cast-iron guide columns,

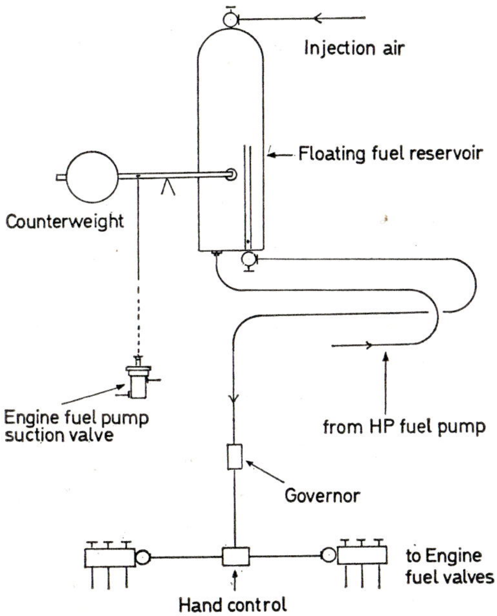

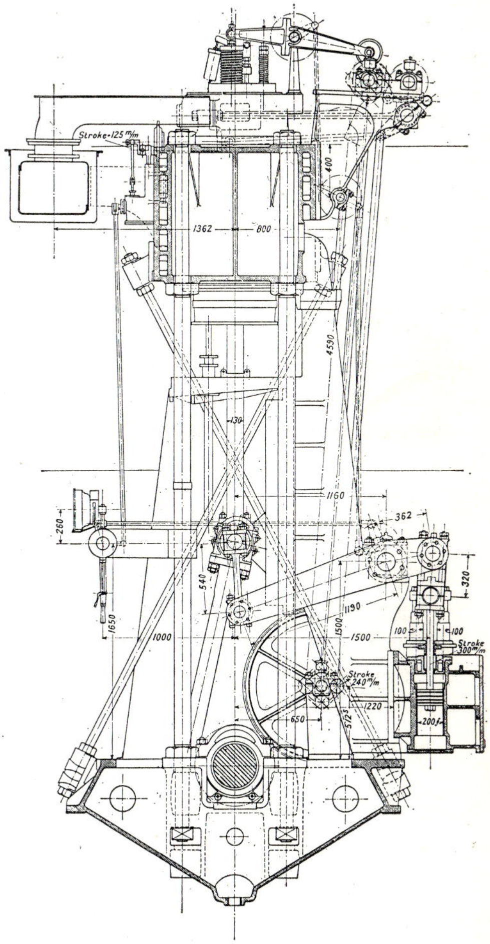

60 *Left:* M/V *Vulcanus* (1910)—Integrated fuel and injection air system at cylinder top level
61 *Right:* M/V *Juno* (1912)—Section through main engine open framework, showing camshaft drive and double camshaft

steel tension columns and steel diagonal ties. There were six heavy-cast columns at the back of the engine, one for each cylinder, taking the dead load and supporting the guide plate. A steel column was then placed at each corner of each cylinder, sixteen all told, to take the load, and eight diagonal tie bolts braced the whole engine structure together, the crankcase, of course, being open for drip lubrication. It should be noted that the cylinders only rested on the iron columns, they were not bolted to them, the flanges not being drilled.

The running gear was oiled by sight-feed lubrication, the waste oil falling into the oil-tight sump being pumped up to a tank to be filtered and used again, and the sea-water cooling pump was driven off the crosshead by beams.

Camshaft drive and reversing were similar in principle to those of the *Vulcanus*, but more sophisticated as a result of experience. The crankshaft drove a layshaft through spur gearing at 2:1 and this layshaft rotated the camshaft direct through four cranks and coupling rods. As in the earlier ship, there were two camshafts, one for ahead running, the other for astern, but in this case they were contained in a housing that slid bodily from port to starboard, thus bringing the reverse cams under the rocker rollers. This lateral movement of the camshaft housing on a flat surface led to minor alignment difficulties. When reversing, the coupling rods, acting as radii round their crankpins, would require their upper ends to follow an arc and not a flat chord, and this slight deviation from the theoretical path tended to lift the coupling-rod brasses momentarily off their crankpins. This, however, was dealt with by having half-brasses only, and keeping these rods in tension at all times.

In later models, Werkspoor adopted a different method of altering the valve timing for astern running. The single camshaft was equipped with ahead and astern cams, fitted side by side, for each working valve, and the valve rocking levers were mounted on skew eccentrics keyed to a layshaft that could be rotated by the reversing gear. When reversing, the rotation of the layshaft lifted the rocker rollers clear of the ahead cams and brought them down in alignment with the astern cams.

In the *Juno* the blast-air compressor was driven off the No. 5 cylinder crosshead, the cast-iron column acting as an intercooler with cooling coils in the bore, and, to complete the picture, a separate diesel engine developing 100 HP on two cylinders, drove a three-stage air compressor, a cargo pump through a belt drive, and independent bilge and circulating pumps. The modern note was introduced by cooling the pistons by fresh water, and by having the bridge-engine-room telegraph incorporated with the main engine controls.

Now although the descriptions of these earliest marine diesel engines might lead one to suspect that motorships in 1911 were as rare as nuclear-powered ships in the 1960's this view would be quite misleading. In a lecture given by Dr Diesel in 1912 he quoted the following list of motor-driven vessels, built or under construction, at the end of 1911:

Tankers
12 Russian on Volga and Caspian, up to 1200 HP.
4 German largest at 4200 HP.
1 British, at 800 HP.
1 USA on Great Lakes.

Tugs
22 Russian on Volga and Caspian.
3 German.
10 other countries.

Auxiliary sailing vessels
8 various countries.

Merchant ships
9 Russian on Volga and Caspian.
7 German.
7 Scandinavian.
4 Dutch.
3 British.
2 Swiss.
2 Congolese.

He also included 20 to 30 warships, 115 submarines, 14 fishing vessels and 16 miscellaneous craft.

It will be noted that the motor ship for inland waterways was pioneered by the Russians, no doubt due to the large quantities of oil fuel available in the Caspian refineries. The pioneer motor-ships, like the pioneer steamships before her, depended on adequate fuel supplies being available along the contemplated trade routes.

In 1912 the motor ship *Eavestone* was commissioned and fitted out with a four-cylinder two-stroke-cycle engine built jointly by Carels Frères of Ghent, who supplied the diesel cylinders, valves and camshaft gearing, and Richardsons Westgarth & Co. Ltd, who provided the rest of the machinery, virtually a plain open-frame reciprocating engine similar to the conventional steam engine; for marine engineers were still not too happy to have the running gear hidden inside a casing. This engine had four cylinders 20 inches bore by 36 inches stroke and developed 800 BHP at 95 rpm. Two scavenge-air pumps were placed at the back of the framework, driven by beam levers from Nos. 2 and 3 crossheads in much the same fashion as the air-pump in a steam engine, and it was calculated

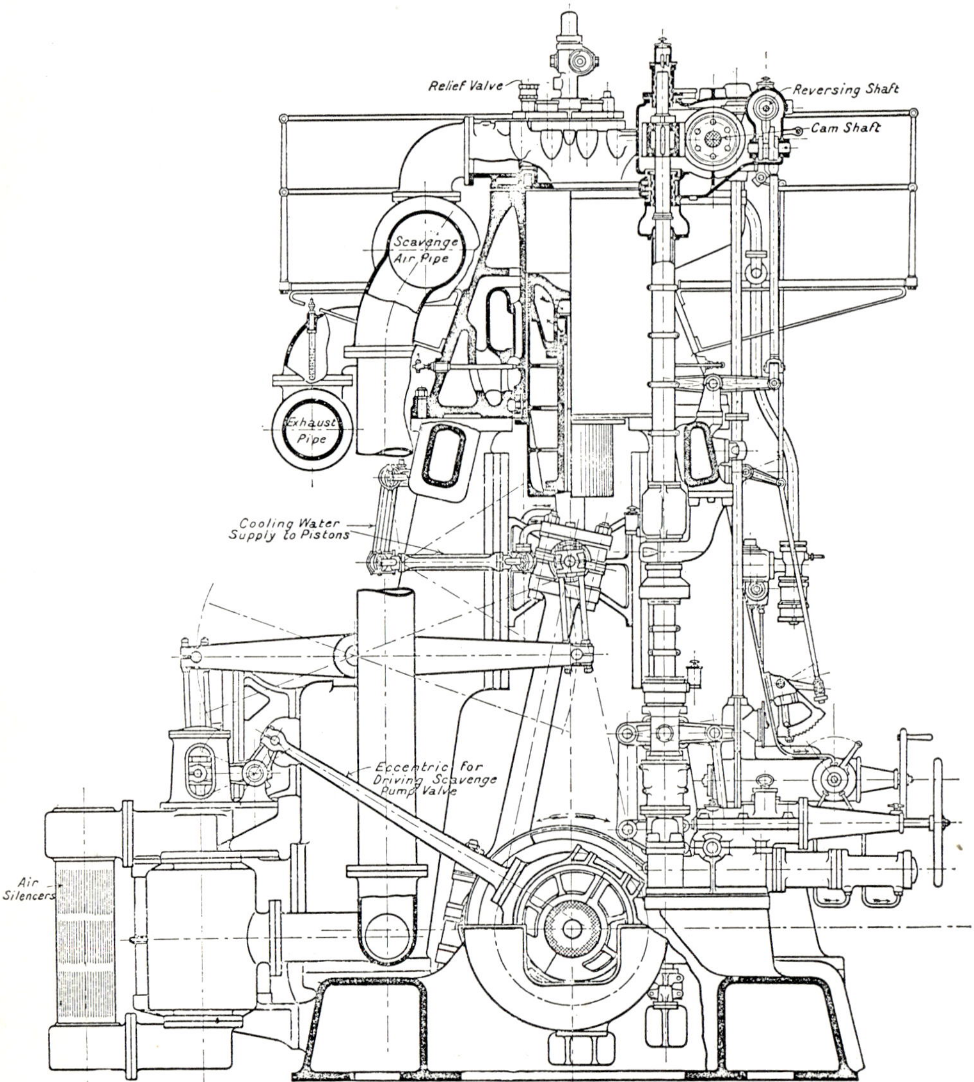

62 M/V *Eavestone* (1912)—Section through main engine showing scavenge pump valve eccentric drive and compound reversing gear arrangement

that with one pump put of action, the other would supply sufficient scavenge air to enable the engine to run at slow speed. The pump valves were mechanically operated by means of a large eccentric rod working on an eccentric on the shaft,

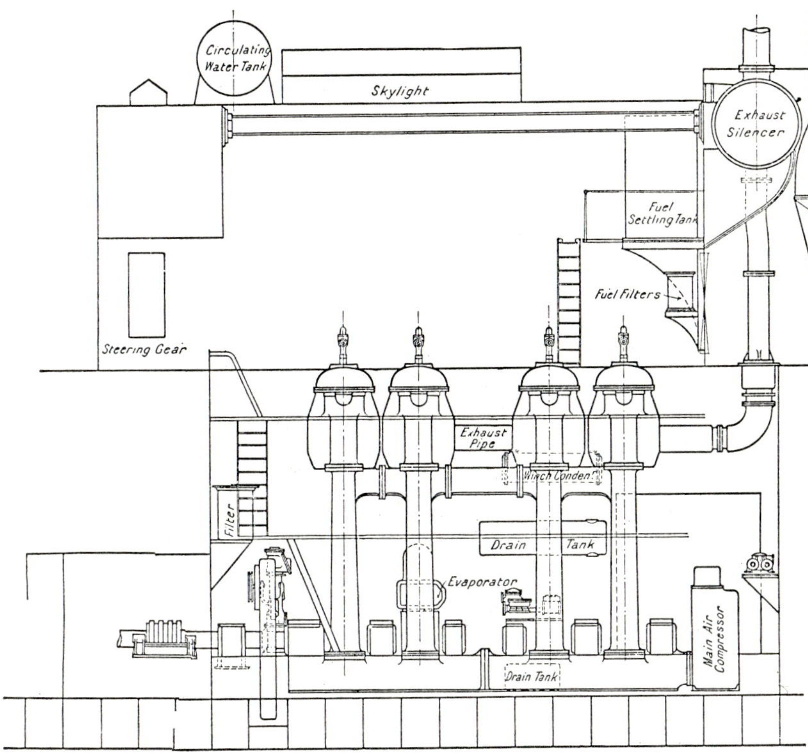

63 M/V *Eavestone* (1912)—Elevation of engine room

although in later engines of this type, automatic pump valves were fitted, but it is worthy of note that even at this early date it had been found neccessary to supply relief valves at each end of the scavenge manifold. These scavenge-pump levers also worked the cooling water, bilge and fuel transfer pumps.

Sea-water cooling was used for the pistons, cylinder liners and heads, guide plates and main bearings. The system of piston cooling compared very well with modern methods. The piston was built up of a heavy heat- and pressure-resisting hollow casting fitted to the piston rod by a bolted flange, and a long hollow skirt to blank the exhaust ports. The piston rod was hollow bored, with an internal pipe for the water supply, the return being along the annular space between the internal pipe and the piston-rod bore. There were no telescopic pipes, the water supply and return being by swinging links between the water main on the engine columns and the distribution chest on each crosshead.

Unlike the two-stroke engine of the day, where the cylinder liners had both

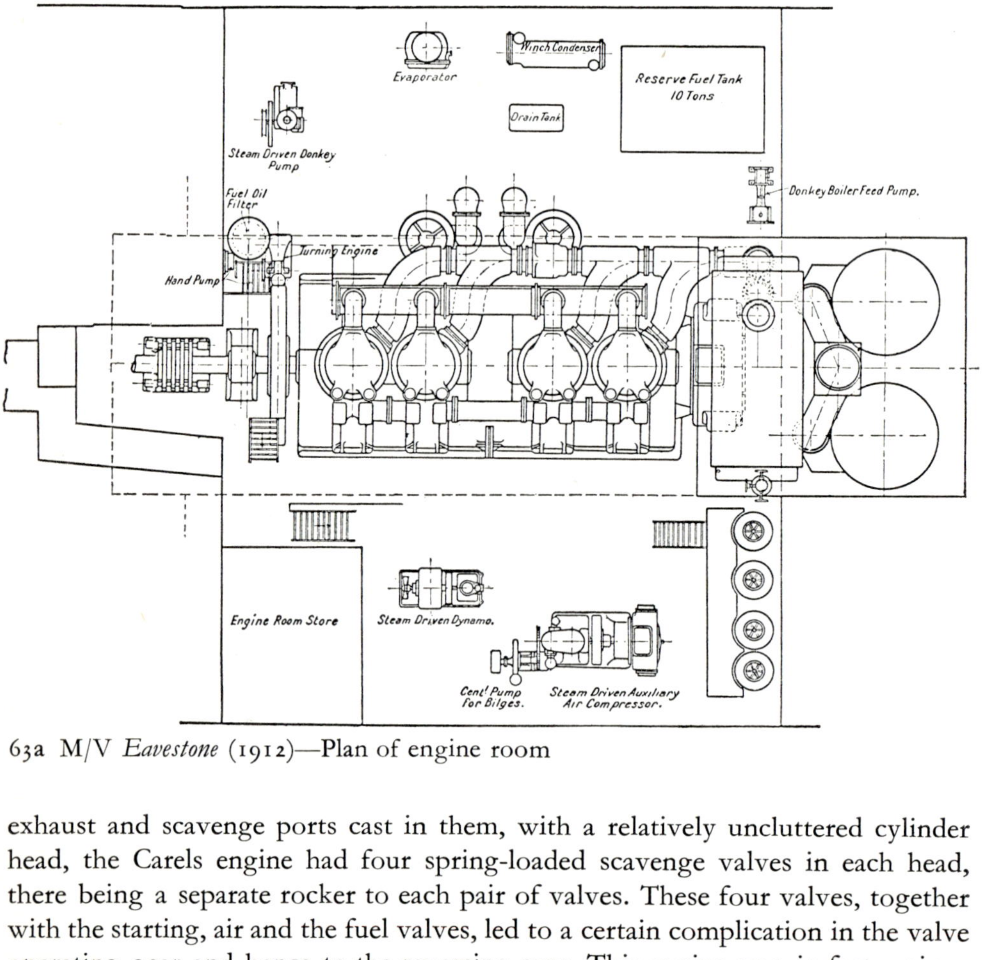

63a M/V *Eavestone* (1912)—Plan of engine room

exhaust and scavenge ports cast in them, with a relatively uncluttered cylinder head, the Carels engine had four spring-loaded scavenge valves in each head, there being a separate rocker to each pair of valves. These four valves, together with the starting, air and the fuel valves, led to a certain complication in the valve operating gear and hence to the reversing gear. This engine was, in fact, unique in having a compound-reversing mechanism involving two different methods of changing the valve timing, (a) by altering the camshaft phasing relative to the crankshaft and (b) by using ahead and astern cams, and it will be appreciated that method (a) was applicable only to two-stroke engines.

In the early Carels engine the timing difference in the scavenge valves between ahead and astern running was about 35°, and as the camshaft drive was by spiral gearing from the crankshaft, a partial rotation of the camshaft relative to the crankshaft was obtained by lifting the vertical drive shaft bodily. This operation altered the scavenge-valve timing without the necessity for lifting the valve levers

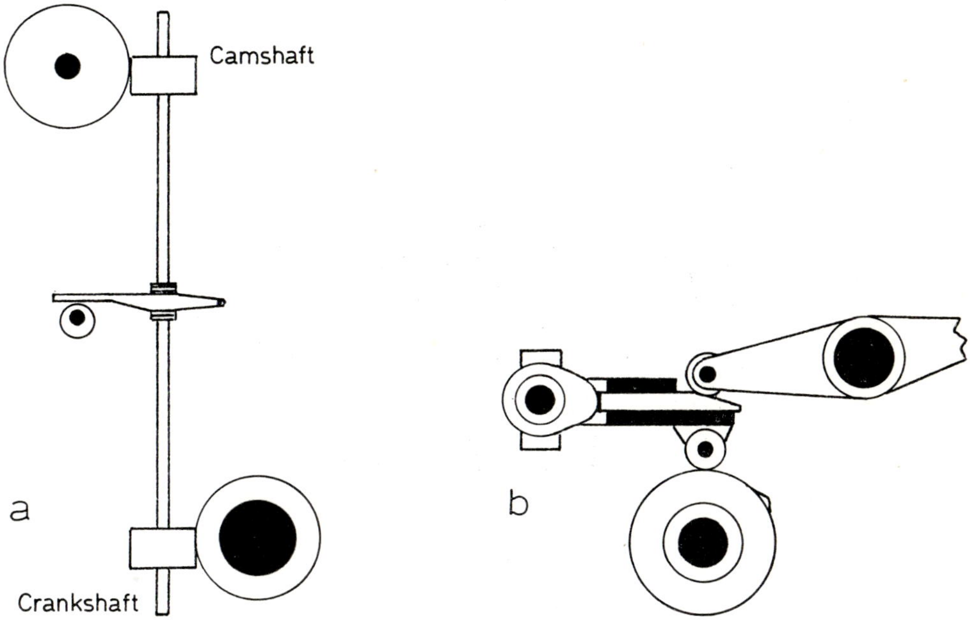

64 M/V *Eavestone* (1912)—Compound reversing gear:
(a) by lifting camshaft vertical drive shaft for scavenge timing
(b) by selective roller wedge action for fuel and starting cams, ahead or astern

off their cams, but it was not suitable for the starting-air and fuel-valve settings. In this case, reversal was obtained by fitting duplicate cams, allowing for the effect of partial rotation of the camshaft in siting the astern cams.

The camshaft was therefore made with ahead and astern cams for starting and fuel valves only, and their rockers were fitted with rollers wide enough to span both cams. A selective roller wedge piece was inserted between the cam and the valve rocker, mounted on a light manœuvring shaft with the wedges controlled by small cams. By rotating the manœuvring shaft, each small cam withdrew its wedge from below the valve rocker and gave sufficient clearance between cam and roller to allow the wedge piece to be moved longitudinally into astern position, the movement of the manœuvring cam thereupon pushing the wedge between the roller and the astern cam.

The manœuvring mechanism was incorporated into that for camshaft reversing by using the selective wedges to slide under their respective rollers in correct

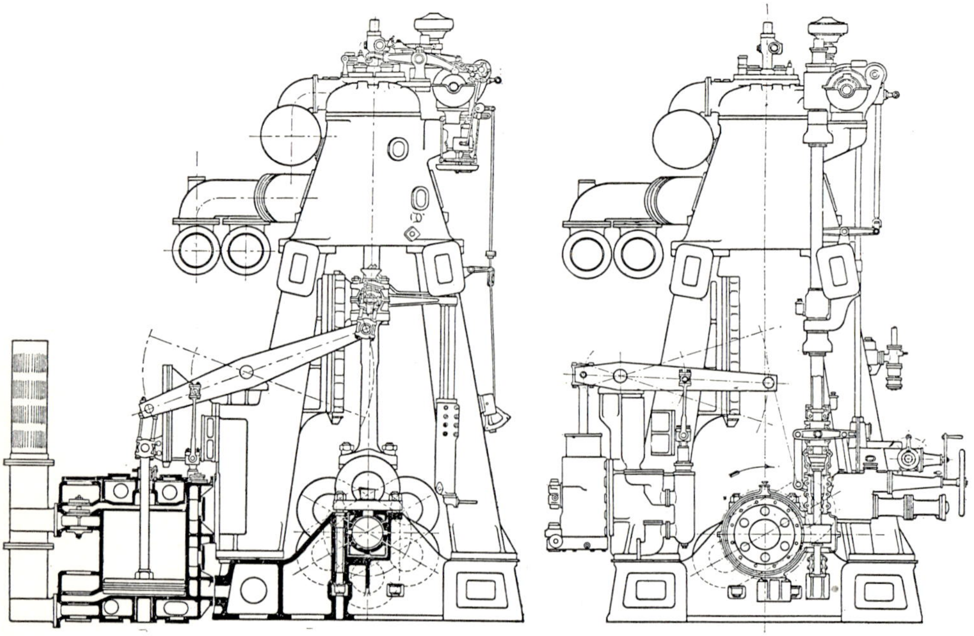

65 M/V *Rolandseck* (1912)—End elevations of main engine, showing scavenge pump drive and compound reversing gear

starting sequence: first all cylinders on air, then half on fuel, half on air and finally all on fuel.

Injection air was supplied by a three-stage Reavell radial compressor driven off the forward end of the crankshaft, there being in addition a smaller compressor driven independently by a steam engine.

In general the arrangement of the *Eavestone*'s engine room was similar to that of a contemporary tramp, even to the banjo-type ballast donkey pump which also served as a turning gear. These steam auxiliaries were supplied with steam by two vertical coal-fired donkey boilers.

Although her engine room may have looked like an ordinary steam engine room, there was no mistaking the early 'diesel look' in *Eavestone*'s outward appearance. The main engine exhaust pipe was carried up one of the derrick posts on the bridge deck, which was additionally graced by the distinctive stove-pipe funnel used for the two donkey boilers.

A slightly more sophisticated version of the Carels engine could be seen in the motor ship *Rolandseck*, built in 1912 by A. G. Tecklenborg of Geestemunde. This engine had six cylinders 20 inches bore by 36 inches stroke but was similar to the *Eavestone*'s engine in other respects, being open framed for hand lubrication. Also, like the British-built engine, the crossheads were provided with high-pressure lubricating oil pumps worked by the swing of the connecting rods. In this ship, however, the engine exhaust gases were passed through a donkey boiler.

An extra feature in this installation was the extensive network of compressed-air equipment, there being no less than six compressors installed. The main blast-injection compressor consisted of a four-cylinder three-stage Reavell unit driven off the end of the crankshaft, while a service two-stage compressor was worked off No. 2 cylinder crosshead. A 100 HP auxiliary diesel engine drove its own injection-air compressor plus a manœuvring compressor, while a small diesel lighting set also required its own blast-air supply. Finally, there was a steam-driven emergency compressor in a corner for good measure. The service compressor requires some explanation; it was used solely for the steering engine after 'Full Away' and the working pressure was kept to 100 lb, it being possible to cross-connect one of the starting-air bottles to the system in an emergency. Whilst manœuvring, the steering gear took steam from the donkey boiler.

It will be appreciated that up to this period, the diesel engines just described, together with their contemporaries, represented the first faltering steps of the marine engineer to introduce this new type of engine into the marine field. These engineers, shipbuilders and owners had been presented with a completely new concept of prime mover but had not yet developed a medium in which to use it to advantage, and the history of the marine oil engine runs parallel with that of the petrol motor which had perforce to be built into a horseless carriage designed along the same lines as its horse-drawn prototype. The principle of compression ignition was built into a steam reciprocating engine of a type with which everyone had long been familiar, and the engine room was still a steam engine room with steam-driven auxiliaries that everybody knew and understood.

The *Selandia* changed all that. Built by Burmeister and Wain of Copenhagen in 1912 for Det. Ostasiatiske Kompagni, she may claim to be the pioneer motor vessel in the sense in which we understand the term, a vessel 370 feet long, 53·2 feet beam and 27 feet depth developing 2500 BHP on twin screws at 140 rpm giving a speed of 11 knots. She was also the pioneer for all-electric auxiliaries, the donkey boiler being there as an emergency unit only.

Each main engine, working on the four-stroke cycle, had eight cylinders $20\frac{7}{8}$ inches bore by $28\frac{3}{8}$ inches stroke, and apart from minor details the engines differed surprisingly little from those of the same make built right up to the end of the blast-injection period, about 1928. These engines were not self-contained as in the later models fitted with a three-stage high-pressure injection-air compressor, and indeed their operation depended on an independent source of air for this purpose. The layout of the engine room was fairly simple, there being the two main engines occupying the full length of the machinery space, with a four-cylinder auxiliary diesel engine in each wing, port and starboard. These two auxiliaries drove their own blast compressors and in addition were coupled to a dynamo and an independent two-stage air compressor delivering to the four large air receivers at about 300 lb pressure, suitable for starting and reversing. Each main engine drove a small high-pressure air compressor from an extension of the crankshaft at the forward end, and when running, this HP compressor drew air at 300 lb from the main receivers and delivered it at up to 1000 lb pressure into the main blast bottles at the engine controls.

Another feature of the *Selandia*'s main engines which was simplified in later models was the rather odd form of camshaft drive very similar to that of the Werkspoor engine. The crankshaft drove a 2:1 idler shaft through spur gearing, this shaft having two small cranks, one on each side of the wheel at right angles to each other. A pair of long connecting rods on the two cranks drove an upper idler shaft through another pair of cranks, and this second idler shaft drove the camshaft at the same speed. Reversing was by the method used in all subsequent long-stroke engines of this make, where the camshaft was placed at mid height and the cams operated the cylinder-head valve rockers through the intermediary of long push rods whose lower ends were held in position by links on a cranked layshaft. The air-hydraulic reversing cylinder worked a rack which meshed with a spur wheel on this layshaft, there being a slanting groove on the back of the rack which engaged a projection in the camshaft thrust block. In reversing, the rack moved upwards and turned the layshaft one complete revolution, the cranks removing the rollers clear of the cams and replacing them in the same position; meanwhile the slanting groove in its back had moved the camshaft in a fore and aft direction to bring the astern cams below the push-rod rollers.

The *Selandia*'s engine room, being completely electrified, released the engine-room watchkeeper from the tyranny of steam, of multiple boilers with their concomitant stand-by losses, and of stokehold crews, soot and ash disposal, choked bilges and leaky glands inherent in steam plant, and enabled him to concentrate on the machinery proper. The drawback to electrical auxiliary power, however, was that it demanded a certain professional training not possessed by the marine engineer of the day, and the engine-room staff, although considerably reduced by the removal of the black squad, had to be reinforced by electrical engineers.

The *Selandia*, having pointed the way and set the fashion for the motor ship of the future, driven by long-stroke crosshead-type enclosed, direct-reversible oil engines, with all-electric auxiliaries, was followed in rapid succession by two sister ships and a host of cargo and passenger vessels of similar size equipped with a great variety of machinery. Famous engineering firms throughout Europe that had been building oil and gas engines for land stations over the years started producing marine engines either of their own design or under licence from makers of an established type.

Although the motor ship was a success, even in those pioneering years before the 1914–18 war, this success was not immediate and unqualified. The principal merit of the diesel engine lies in its fuel economy, although this was not so apparent then as it became in the 1930's, when oil-burning turbine steamers showed a figure of 1 lb/IHP/hour against the diesel motor's 0·35 lb/IHP/hour. In later years, in the 1950's and 1960's, when much more sophisticated steam machinery was introduced, this figure dropped to 0·60 lb/IHP/hour in really big steamers. But in the early years of the twentieth century relative economies were not as simple as all that. The average freighter had to earn its living all over the globe, while fuel suitable for marine oil engines was available only in the larger bunkering ports, and very expensive at that.

Around 1912 the cost of steam coal was 12*s*. a ton, while diesel oil cost up to 45*s*. a ton. Thus, for an average-sized tramp steamer of, say, 1000 SHP, with a consumption of 1·5 lb/IHP/hour steam or 0·36 lb/IHP/hour diesel power, the daily fuel bill would work out at £8.8*s*. for coal or £9.14*s*. for oil, a daily net loss of £1.6*s*. debited to the motor ship. This figure was increased by the necessity for carrying extra engineers, oilers and, possibly, electricians, and in those early years, when motor ships were considered unhealthy (they still are), diesel engineers were paid at a higher rate than steam engineers.

These costs were more than offset by the enormous increase in cargo space available through the elimination of the boiler room and bunkers with their attendant fireman and trimmers, and the almost total absence of stand-by losses.

Surprisingly, these early marine diesels in general gave a very good account of themselves and their official log-books seldom mention break-downs that were peculiar to the diesel cycle. The engines were massively built, after the fashion of the steam reciprocators, and although they were subject to many stoppages for repairs to minor components, they generally managed to get to port under their own power. In my own modest experience in motor ships, an average of three stops a week for valve renewals was considered good going, and on separate occasions we brought the ship safely home with a broken main engine 'A' frame and a main switchboard that caught fire. If diesel engineers were paid more than their steam colleagues, it was because they earned it.

The earliest motor vessels were really trying out the diesel engine under sea-going conditions which are completely different from those on land, especially as regards electrical machinery, and any faults noted by the crew were rectified on return to the home port. They were, in fact, guinea pigs, and it might be of interest to pass in review the first crucial years of these pioneer motor ships to see how the diesel principle behaved at sea. One of the earliest, the *Romagna*, an Italian vessel built at Ancona in 1910, had an engine supplied by Sulzer Brothers, of Winterthur. After commissioning, her compressors were found to be inadequate and she was supplied with larger compressors. She was lost at sea in 1911.

The *Vulcanus*, already described, was quite successful apart from a burst air receiver, and made a name for herself out in the Dutch East Indies. The *Toiler*, British built with a Swedish engine, ended up in service on the Great Lakes, but after a while her owners decided she was underpowered, the engine having eight cylinders $9\frac{7}{8}$ inches bore by $14\frac{1}{2}$ inches stroke, which, with the brake mean effective pressures of those days must have given quite a low power for a vessel of 1660 tons gross. There was thus no complaint against the diesel engine as such. The *Calgary*, a similar vessel to the *Toiler*, built in 1912, for the same service, was given a much more powerful engine with cylinders $11\frac{1}{2} \times 17$ inches bore-stroke.

The *Selandia* had cylinder trouble on her first voyage out East with two seized pistons, and on return had her exhaust valves replaced by new valves of improved design. Her sister vessel *Christian X* had fuel-valve and cylinder-liner trouble, the latter attributed to scale formation from sea-water cooling. Another similar vessel, the *Suecia*, had the not-uncommon accident of a blast-injection receiver bursting through oil contamination of the compressed air and a blow-back from the intercooler.

The *Fionia*, a twin-screw motor ship with the largest bore diesel cylinders of her day (29 inches), had serious compressor trouble on her maiden voyage and was compelled to return to port to have these renewed of a different pattern.

The *Eavestone* was not so fortunate, as in 1913 she was forced to put into the Azores with her machinery disabled and had to remain there until new pistons and cylinder heads could be sent out to enable her to complete the voyage. A similar occurrence overtook the *Monte Penedo* at about the same time. This vessel was fitted with engines supplied by Sulzer Brothers, and on her maiden voyage she arrived at a South American port with serious piston trouble and had to await the delivery of new pistons sent out from the engine-builders. Both these engines and the Carels engine of the *Eavestone* ran on the two-cycle principle, although the *Rolandseck*'s engine, also a Carels engine, gave no trouble at all.

The *Juno*'s probationary years were trouble-free, as were the *Emanuel Noble*'s and the *Loudon*'s. The only other vessel worthy of mention in this review of 1912 vintage motor ships was the *Fordonian*, whose British-built Carel's engines were fitted with cast-steel cylinder heads. For some reason not readily understood, these heads were replaced by cast-iron ones, not because of any defects in them but as a result of experience with other engines.

Finally the *Hagen*, built in 1913 with Krupp-type two-stroke-cycle engines on twin shafts, had serious trouble with her compressors, and on return from the maiden voyage was laid up for three months having one of her compressors renewed. This compressor trouble was not peculiar to any one type of diesel engine, it was inherent in all marine oil engines, and inseparable from the blast-injection principle. High-pressure air was used to start and reverse the diesel engine and very high-pressure air was necessary to atomise the oil in the cylinders, and it was the simplest and quickest thing in the world to lose the complete charge of air in a blast bottle by manipulating a wrong valve or even through leakage from a joint or gland. Compressors were the bugbear of the early diesel engine: a good compressor was at best a trouble-maker, a poor compressor was a nightmare, and one of the greatest steps forward made by oil engineers was the introduction of airless injection for marine engines between 1928 and 1930. In the old blast-air system, the fuel-lift pumps, worked from the main engine, simply supplied the oil to the fuel valves at a pressure slightly above that of the blast-air pressure, and it was this air that literally blasted the oil through the atomiser discs and into the cylinder when the fuel valve opened. This valve timing thus required a special cam and valve rocker.

With airless, or 'solid', injection, the fuel is injected direct into the cylinder at very high pressures, the jerk pump itself being timed while the fuel valve is automatic and acts in effect as a relief valve to the pump. But the diesel engine is not self-starting, it still depends on an outside source of energy, usually compressed air, to run it up to speed.

Earlier on in this chapter the marine oil engine had been compared to the motor-car engine on the score that they both appeared on the scene at about the same time and were both used as extensions of the earlier forms of propulsion they displaced: steam for one, the horse for the other. But there the resemblance ended. The automobile engine, even in its early years, was the same basic model in nearly every type of car, irrespective of build, and the only difference between one make and another lay in the number and disposition of the accessories. The highly developed and sophisticated petrol engine of today is essentially the same for all modern cars, and has not changed greatly in design since the turn of the century.

Not so the diesel engine, however. Even in its early years the marine oil engine was fairly equally divided between the four-stroke and the two-stroke principle, and as engine-builders acquired more experience and consequently more confidence, these two main types were subdivided into single- and double-acting, with, at a certain period, the added choice of blast or airless injection of fuel. Again, two-stroke engines could have their scavenge air supplied by independent blowers driven by an auxiliary engine, or from a turbo-compressor driven by the main engine exhaust gas, from a reciprocating pump worked off the beams, or yet again from the underside of the main pistons.

The engines themselves could be of the crosshead or trunk-piston type, or, if preferred, built on the opposed-piston principle with several makes to choose from. Finally, if the ship-owner still could not make up his mind when faced with this choice, he could have an engine working on both steam and diesel cycles simultaneously. The marine oil engine was an individual engine in that every type was completely different from every other, while any one builder could change and remodel his engine several times over while still somehow retaining the essential features in each case.

Some of the better-known makes of early marine oil engine are briefly described below; as this is a history of the diesel engine in general and not a comprehensive treatise on each of the different types, only the salient features of each engine will be dealt with and no attempt will be made to carry the description into the present-day models. Also, it should be borne in mind that many patented engines were and still are built under licence by engine works in different parts of the world and thus reflect an approach in design not always evident in the prototype.

The Werkspoor engine has already been described in its early applications to the *Vulcanus* and *Juno*, and its principal characteristics were maintained for as long as blast injection was used. A tall, four-stroke engine with the camshaft at cylinder-top level, driven by four connecting rods from a crankshaft-driven layshaft. The valve rockers were mounted on eccentrics set askew upon the fulcrum shaft, and the valve ends of the rockers were prevented from lifting by means of a stopper bar. Upon reversal, the fulcrum shaft was rotated one half revolution thus lifting the rocker rollers off the ahead cams and, by means of the skew effect, placing them on to the astern cams.

The integrated fuel/blast air-control mechanism, already described, was also a feature of this type, and engines with this method of fuel control were still in existence in the 1950's. These long-stroke engines with blast injection, as indeed most large engines of the period, were extremely flexible and could run smoothly at 30 rpm provided the blast pressure could be maintained.

In the double-acting Werkspoor engine, with valves at top and bottom of each cylinder, reversing was carried out by fore-and-aft movement of the camshaft, the rockers being mounted on eccentrics placed fair on the fulcrum shaft. Like most double-acting engines the firing pressure below the main piston was lower than that above it, and starting air was introduced at the top only, the bottom exhaust valves being held open during this manœuvre until the engine was firing on all cylinder tops.

The Burmeister and Wain engine, originally designed in Copenhagen but now built under licence throughout the world, was also a four-stroke unit, made either in the long-stroke crosshead type or in a shortened, trunk-piston version for cross-channel and special-type ships. In the crosshead type the camshaft was at mid height, driven by a spur-wheel train or later by roller chain, and the valves were operated by long push rods mounted on a reversing crankshaft parallel with and adjacent to the camshaft. When reversing, an air-driven rack meshing with a spur wheel on the reversing shaft gave it one complete revolution, lifting the cam rollers off their cams for a short period. Meanwhile, a collar on the camshaft, engaging in a diagonal slot in the back of the rack, moved the camshaft into the astern position.

The fuel valve on the blast-air-injection engine, contrary to normal marine practice, was of the mushroom type, opening into the cylinder, and therefore required a very careful handling of the air-pressure control when running at slow speeds, as the period of fuel-valve opening was regulated by the cam only, and was independent of the amount of fuel delivered by the pump. Thus, at 120 rpm, the fuel valves would remain open for, say, $\frac{1}{16}$ second per firing stroke only, while at 30 rpm this period would extend to $\frac{1}{4}$ second. A sudden call for 'Dead Slow Ahead' in fog or ice could easily catch the watch keeper off balance, and unless he had his wits about him he would run short of blast pressure and his engine would falter to a noisy and spectacular stop. To prevent this, the fuel rockers were mounted on eccentrics on the fulcrum shaft whereby the valve lift could be reduced for slow running.

The double-acting version of the four-stroke Burmeister and Wain engine was fairly successful in its way and was to be found in many famous passenger and cargo-liner fleets, the largest being the *Britannic* and the *Georgic*, each of which had two ten-cylinder engines in the main engine room, with four six-cylinder diesels driving dynamos and four four-cylinder diesels driving compressors for the main engine blast-air system.

The arrangement of the top cylinder valves was similar to that of the single-acting engine, with long push rods from the camshaft at mid height, while the bottom valves were worked by short push rods extending below the camshaft. Reversing was on the same principle, there being two reversing shafts, one for each cylinder end.

This engine, as indeed any double-acting blast-injection engine, required a highly trained and devoted engine-room staff and I have yet to meet a marine engineer who regretted their passing.

In the 1930's a two-stroke-cycle version of the Burmeister and Wain engine appeared, single-acting at first, then as a double-acting engine. In the single-acting type the exhaust was by a large poppet valve set squarely in the centre of the cylinder head, and scavenge-air supply, from rotary blowers driven off the camshaft chain drive, was introduced through ports in the cylinder liner.

In the double-acting type the exhaust gases were discharged through piston valves in the centre of each cover, the lower cover piston valve also carrying the piston-rod gland. These piston valves, top and bottom, were yoked together and were driven originally by long connecting rods coupled to a layshaft at the back of the engine geared to the crankshaft. In later models the exhaust pistons were driven by large eccentrics fitted directly on to the crankshaft—in practice these exhaust valves acted as secondary pistons and contributed towards the main engine power. Finally, the exhaust pistons eventually were made as full-sized pistons, there being for each working cylinder one main piston driving the crank through piston rod, crosshead and connecting rod, and two secondary pistons, one above and one below the main piston, driving the shaft through a pair of large eccentrics, one on each side of the crank, the ratio of main to exhaust piston stroke being roughly 3:1.

Scavenge air was supplied by mechanically driven blowers or by exhaust-gas-driven turbo-compressors.

Another four-stroke engine, the Vickers, was outstanding in that it used air-less injection right from the start, fuel being supplied at up to 4000 lb pressure, according to speed, by a set of pumps driven from a small crankshaft at the front of the engine. The fuel pumps all delivered into a common rail or pipe at the back of the engine from which a lead was taken to each fuel valve. The fuel valves were worked by cams on the camshaft at cylinder-top level, i.e. they were of the mechanical-injection type. The inlet and exhaust valves on the cylinder head had their individual bellcrank rockers mounted on the valve casings, and these rockers were in turn connected to another set of bellcrank rockers on the reversing shaft above the camshaft. Reversing was by rotating the reversing shaft and sliding the camshaft.

Salient features of the Vickers engine were the cylinder head and the exhaust valve. The cylinder head was an open cast-iron box with valve pockets around which sea water circulated, with a forged-steel flame plate to take the combustion stresses. The exhaust valves were made hollow for sea-water circulation, there being a steel tube fitted inside the stem to direct the water on to the valve head, the return taking place between this pipe and the valve stem, while armoured rubber hoses connected the moving valves to the supply and return manifolds. This system was good in theory but when the exhaust valves burnt through, as they not infrequently did, and blew the hoses off, the affected cylinder had to be cut out at once.

The Beardmore-Tosi engine, also a four-stroke unit, was fitted into several early motor ships and it had several characteristic features not found in other engine types. It was built on the désaxé system, whereby the centre line of the cylinders is offset by about 2½ inches from that of the crankshaft, the offset being positive in running ahead. The advantages claimed for this design were that with the crank at top centre, the piston was already over the top and on its way down, it had a quite perceptible lead and therefore a smoother impulse on the firing stroke. The length of the connecting rod was also slightly reduced by this offset and hence the overall engine height, and finally the claim was made that the guide pressures were considerably reduced.

The inlet and the exhaust functions were 'twinned' on this engine, there being two combined inlet-exhaust valves per head, both identical, and both opening simultaneously for fairly long periods of stroke. These valves were timed by a pair of director flaps in the cylinder head. The inlet-exhaust valves were open during the exhaust stroke when the director valves guided the gases into the exhaust manifold, and remained open during the suction stroke, during which time the director valves, having swung round in the valve passage, guided the air from the induction manifold into the cylinder. The director valves were driven by eccentrics on the camshaft through poker rods, and the aim of this added complication was to cool the exhaust spindles by good clean air. This air, incidentally, was drawn from the lantern framework below the open cylinders and if it wasn't always clinically pure, at least it ensured that those noisome blows past the piston rings did not pollute the engine-room atmosphere.

Reversing was simple. The valve rockers, mounted on eccentrics on the fulcrum shaft, had wide fork ends allowing a generous lateral movement of the rollers. The camshaft carried the usual double cams, ahead and astern, and when putting the engine into reverse position, the fulcrum shaft was revolved by an electric motor through gearing, to lift the rollers clear of the cams, while a light hand-reversing shaft slid the rollers along their pins into position above the astern cams.

Reversing in most of the early diesels was achieved by a dual movement of the control gear, lifting the rocker rollers off the cams and sliding the camshaft in a fore-and-aft direction. In the North British four-stroke engine, it was the camshaft that dropped away from the rockers, the bearings being carried on eccentrics mounted on a reversing shaft fitted below the camshaft.

The McIntosh and Seymour engine, a four-stroke air-injection engine similar in principle to those described above, had no special features of note other than the fuel-pump drive. Here, the pump for each cylinder was driven by its eccentric sheave, and all the eccentric sheaves were in turn mounted on an inclined eccentric sleeve on the pump shaft. By manipulating the position of the inclined eccentric relative to the plunger eccentrics, the pump stroke could be controlled from maximum to nil delivery.

Another four-cycle long-stroke engine was that built by Maschinenfabrik Augsburg Nurnberg A.G. (MAN). This engine was unusual in that it could burn boiler oil or tar oil, there being two sets of fuel pumps, one set being provided to inject a small quantity of light oil (about 5% of the charge) to ignite the main charge.

The Deutsche Werke engine had an odd resemblance to the old steam reciprocator in that the cylinders were supplied with front and back feet bolted solidly to the cylinder columns. The camshaft was driven by the crankshaft half-time shaft through four long connecting rods in a fashion similar to that of the Werkspoor engine, and the injector-air compressors, instead of being at the forward end of the crankshaft, were driven off crosshead beams.

There were many famous makes of marine motor working on the two-stroke cycle, but as in the case of their four-stroke counterparts, not every type survived its formative years: it took not only capital and good design to push a new engine on the market, it required a great deal of luck.

Sulzer Brothers, of Winterthur, were among the earliest of marine diesel builders, and it may be recalled that they were the first to introduce direct reversing. The engines used port scavenging and port exhaust, the cyclinder head carrying a single valve box centrally placed which contained the fuel and starting-air valves, both driven by a camshaft placed at cylinder-top level. Reversing was an extremely simple matter and entailed no heavy manœuvring mechanism. The valve rockers were not fitted with rollers at their cam ends, but with short triangular box links, carrying two rollers at their lower ends, each roller being in the same plane as an ahead or astern cam. The lower end of each roller box was attached by a short link to the reversing shaft. By partial rotation of this shaft, the ahead roller was pulled clear of its cam while the astern roller landed on to its own cam, and the engine was in position for starting in reverse.

An unusual feature of the Sulzer cylinder was its timed scavenging. The liner carried a belt of long exhaust ports at the rear and two belts of scavenge ports at the front, the upper scavenge row being slightly higher than the exhaust ports. This upper row was mechanically controlled by a valve or valves and the ports remained shut for the period of exhaust and opened only when the piston had reached the bottom of the stroke; they remained open until the piston was well on to the compressor stroke. The lower scavenge ports were not controlled.

This scavenge valve in its earliest form was a rotary valve driven off the vertical camshaft drive; later it was a piston valve worked direct from the camshaft and eventually it became a series of automatic plate valves attached to the scavenge belt.

For some reason not readily understood, the early Sulzer marine engines were handled from a control station at cylinder-top level: in the tropics a hot and unpleasant situation, requiring a lot of running up and down ladders to attend to the auxiliary machinery and compressors. It is true that the engineer at the controls could glance along the cylinder tops and make sure the camshaft was in the correct running position and that starting-air valves were not sticking, but this seemed an insufficient excuse for the inconvenience caused.

The Neptune engine had an unusual type of cylinder head, consisting of a main cover containing an extension of the cylinder liner, with the cylinder cover sitting on top and containing the fuel and starting valves. The camshaft operating mechanism was after the style of the Werkspoor engine but with added articulation. The crankshaft carried two eccentrics at 90° to each other, each of which carried an eccentric rod giving movement to one end of a bellcrank lever mounted on a horizontal shaft at cylinder-bottom level. The opposite ends of the two bellcranks worked the camshaft through a pair of long connecting rods and cranks.

The early Polar engine was unique in that the scavenge pumps were used to start the engine. The lower part of each cylinder was enclosed to form an under piston scavenge system, supplemented by a reciprocating scavenge pump worked from one of the crossheads. Delivery of scavenge air into the scavenge manifold was through a valve in the distributor box on each cylinder end, both the distributor box and the under-piston scavenge casings being made strong enough to withstand reasonably high-pressure air. To start the engine, the appropriate distributor valve was opened by timing shaft and the under-piston scavenge system acted as a power unit to run the engine up to speed.

Most of the engines already described, both four stroke and two stroke, were eventually made double-acting, and in this connection the Burmeister Wain and the Werkspoor have already been mentioned. Others were the Sulzer, MAN, Deutsche Werke, Beardmore-Tosi, Worthington, Fiat, Hesselman, Krupp and Richardsons Westgarth, etc., some of which failed to make the grade but all of which contributed a lot towards the development of the diesel engine.

Opposed-piston engines have so far not been mentioned, they form a group of their own based on Professor Junkers' patent, and there are or have been a number of variations on this design. This type of engine was probably better known in Germany as the Oechelhauser and in France as the Gobron-Brillié type and it antedates the birth of the compression-ignition engine by a few years. In 1894 a Belgian engineer Pierrez developed an engine using a piston working against a sliding cylinder cover, and in 1903 Sautter, Harlé & Cie of Paris were marketing their horizontal balanced oil engine having opposed pistons in a common cylinder set astride the crankshaft, each piston driving a yoke with a pair of return connecting rods on to two cranks.

One of the earliest marine engines using opposed pistons was a small three-cylinder reversible unit with cranks at 120° working on the two-stroke cycle. Incidentally, the Junkers type is inseparable from the two-stroke cycle, it could be run on the four-stroke principle but would then lose all its advantages of balance and through scavenge. This little engine had cylinders 8 inches bore and a combined stroke of 19 inches and ran at 150 rpm the pistons being of the trunk type and the scavenge pump being incorporated with the blast compressor at the forward end of the crankshaft.

The Junkers engine was made in two forms, simple and tandem, and the cylinders were open at both ends, the combustion being at the middle. The lower piston, through piston rod, crosshead and connecting rod drove the centre crank, while the upper piston, attached to a working beam or yoke, carried two long return connecting rods on to two cranks adjacent to the first one. The two outer cranks, in phase with each other, were at 180° to the centre crank, and each cylinder in the engine was associated with two pistons and three main cranks. The lower piston running gear had to transmit power to its crank directly and was accordingly made of heavier scantlings than the side running gear from the top piston, which was in tension only.

In the tandem type a second cylinder was placed immediately above and in

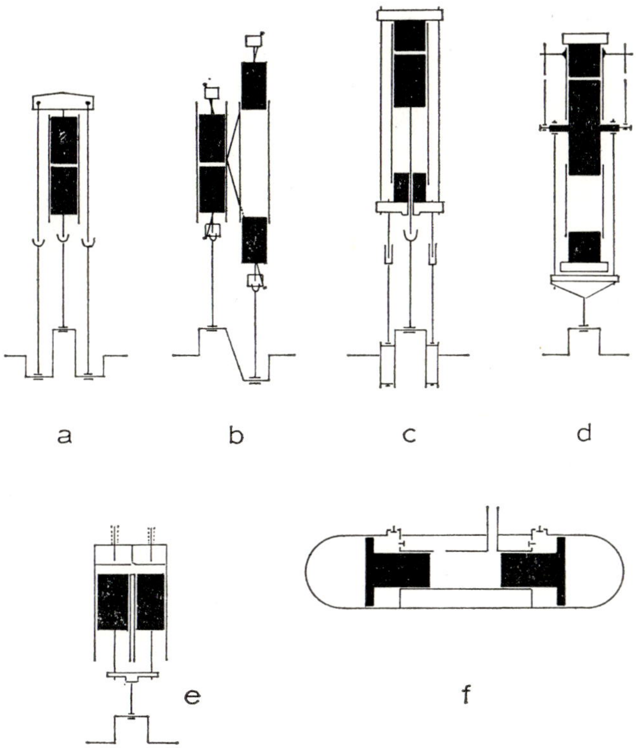

66 Various types of opposed piston diesel engine:
(a) Doxford
(b) Fullagar
(c) Burmeister & Wain—Harland & Wolff
(d) Maclagan
(e) Toussaint
(f) Pescara gas generator

line with the first and their pistons were suitably yoked together to work the three cranks as before, the upper piston of the top cylinder to the lower piston of the bottom cylinder and the lower piston of the top cylinder to the upper piston of the bottom cylinder. So far I have been able to trace only one ship ever fitted out with tandem Junkers engines, a twin-screw installation built in 1912 for the Hamburg America Line, each engine developing 800 BHP on three cylinders, but there is no record of their performance.

The best known of all opposed-piston engines was, and still is, that made by William Doxford & Sons Ltd, of Sunderland, and by the numerous firms under licence. It was also the most popular among marine engineers, possibly because airless injection was used right from the start. This engine differed slightly from the Junkers' original design. The upper piston attached to the transverse beam working in bottle guides on top of the cylinder, drove a pair of side rods attached to their crossheads at the same level as the lower piston crosshead, and connecting rods from these side crossheads lay on to the outer cranks at 180° to the centre crank. As the side running gear was always in tension, both with the piston on the compression and the combustion strokes, they could be made much lighter in section than the centre gear. In the earlier engines the pistons were made of equal stroke, but in later models the upper piston stroke was made shorter than the lower to effect a more favourable balance. The exhaust ports were placed near the top of the cylinder liner, controlled by the upper piston, while the scavenge ports were near the bottom and controlled by the lower piston.

There were two camshafts, one at the front, one at the back, placed roughly at cylinder mid-length, but only the front shaft operated on astern running. Scavenge air was supplied by a large reciprocating pump driven either by the crankshaft or from a crosshead by levers.

The Fullagar engine, also an opposed-piston engine, ran on an entirely different system of running gear from that described above. The cylinder liners were placed in pairs in a common cylinder entablature, there being two pistons working in each liner, and the engine was built up of several such units, while the crankshaft was forged with a pair of cranks at 180° to each other for each cylinder unit. Assuming one cylinder unit with cylinder liners A and B, the bottom pistons in each liner were coupled direct to their respective cranks through conventional crossheads and connecting rods. In addition the bottom piston crosshead in liner A was attached to the upper piston crosshead in liner B by a diagonal rod, and vice versa, the rods being guided for side thrust.

The upper crosshead arrangement was enclosed and formed the scavenge pump for its own cylinder liner, while the crank pit was open, as in the case of the marine steam reciprocating engine.

As already described, the Burmeister and Wain—Harland and Wolff double-acting two-stroke engine works on a modified version of the opposed-piston principle, only in this case there are the main piston working directly on to the crank and a pair of exhaust pistons, yoked together driving the shaft through large eccentrics. For the sceptics, it is just as logical to drive a shaft through an eccentric sheave as by means of a crank, in theory they are the same.

A more unusual type in this series was the Maclagan engine as fitted to the motor ship *Swanley*; it was not an opposed-piston engine in the Junkers sense, but it defies classifying into any other category. This three-cylinder double-acting engine had a single, long piston in each cylinder while the cylinder liner itself reciprocated in phase with the piston, though on a much shorter stroke. The top and bottom cylinder heads were stationary and were secured to cross members of the engine frames while the cylinder itself consisting of two separate lengths coupled together by a pair of tie rods, oscillated between them. Gas tightness between liners and heads was maintained by piston rings. The long piston inside this cylinder was provided with a pair of outside gudgeon pins at mid-length protruding through the opening between the two halves of the liner and which engaged with the fork end bearings of a connecting rod on the crank. The engine crosshead also gave reciprocating motion to the cylinder liner through linkage and rocking levers to a pair of gudgeons on the upper cylinder scavenge belt.

This engine, being double-acting, had a row of exhaust ports and of scavenge ports in each half-liner, the exhaust ports being uncovered by the working piston, the scavenge ports by the movement of the liner past the fixed head, and the scavenge air to and exhaust gases from the cylinder were led from their respective belts to the engine manifolds through telescopic piping.

The fuel and starting valves were fitted in the stationary heads, worked from cams in the ordinary way, and the injection-air compressor was driven off the forward end of the crankshaft, with the scavenge pump at the after end.

Finally there was the Toussaint engine, built in 1912 for use in tugs and river craft: although it did not have opposed pistons it shared all the characteristics of the opposed-piston design. The engine was built up of three-or four-cylinder units working as many cranks. Each unit consisted of twin cylinders set side by side in a single casting and with a common combustion space at the top. Each cylinder had its own cover containing the fuel and starting valves, those in one cover being for ahead running, the other for astern. There was a single trunk piston in each cylinder, the twin pistons being coupled by a single sturdy connecting rod to the crank.

This engine ran on the two-stroke principle with exhaust and scavenge ports at the bottom of each cylinder liner uncovered by the pistons, while the cylinder valves were cam operated in the usual diesel fashion. Although ingenious, the Toussaint engine in effect offered the same power with possibly poorer balance and considerably more complication than an orthodox engine with half the number of cylinders of equivalent area.

This Toussaint engine must not be confused with the Griffin engine which came out about the same time and which bore a very strong resemblance to it in that the cylinders were placed in pairs in the cylinder block, the two trunk pistons being coupled to the one connecting rod. But here the resemblance ended. The Griffin engine ran on the four-stroke cycle, and each cylinder was independent of its twin, with separate valves worked by their own cams. The twin pistons were 360° out of place with each other: when one was on the compression stroke, the other was on the exhaust stroke and there was thus an impulse at every revolution. This Griffin engine could therefore be correctly described as a four-stroke engine working on the two-stroke cycle.

All the compression-ignition oil motors described so far, whatever their shape or working cycle, were pure diesel engines in principle, and they relied on the combustion of oil alone to drive the shaft. The Still cycle, however, introduced in 1924, used a combination of diesel and steam power to actuate the pistons, and it developed along two different paths. In the twin-screw motor vessel *Dolius*, each engine consisted of four cylinders on cranks at 90°, the cylinders being arranged to work on the diesel principle above the pistons and by steam below them. The diesel cycle used the port exhaust and scavenging flow and the simple cylinder-head casting carried only the fuel and relief valves. The steam cycle used the steam expansively first into the aftermost cylinder as the HP stage, exhausting from there into the three remaining cylinders in parallel as the IP stage and from these into an exhaust turbo-blower to provide the scavenge air for the diesel cycle. The inlet and exhaust valves for the steam cycle were oil operated by a distributor shaft, while the top fuel valve was operated by pressure directly from the jerk pump.

In practice, the long piston required no cooling other than that provided by the steam working below it, while the cooling water in the cylinder heads and jackets was used as feed water in the exhaust-gas-heated boiler. This circulating water was kept at about 350°F, and steam was produced at 120 lb. The differential expansion induced by the separate steam and diesel cycles in the one cylinder was allowed for by separate liners for top and bottom.

The aim of this dual cycle was to conserve as much of the heat input as possible, for in engineering, heat is power, and in this the engine was eminently successful. The only loss was the low temperature heat in the funnel gases after they had passed through the exhaust-gas-heated boiler. The fuel consumption figure given for these engines of 0·33 lb/IHP/hour was very low indeed. Also, the mechanical efficiency of the diesel cycle was increased by the absence of the usual blast compressor, scavenge pump or reversing complications, as starting and reversing were by steam direct on to the engine. On the other hand a very low fuel consumption had to be paid for in other ways. In any heat cycle the cost-efficiency curve tends to flatten out more and more at high ratings and the small percentage gain in fuel economy must be bought at enormous capital costs. Without a doubt, the *Dolius* engines were very complicated and as the diesel and steam cycles were inextricably combined, any small defect would naturally affect both cycles.

A modified Still cycle was used in the engines of the twin-screw motor ship *Eurybates* built in 1927. These engines consisted of separate oil motors and steam engines joined together on the one crankshaft. The diesel section consisted of six cylinders 17 inches bore by 45 inches stroke, working on the single-acting two-stroke principle, with two double-acting steam cylinders. The sequence of feed circulation through the cooling-water spaces was similar to the other vessel, and the exhaust-gas boiler supplied steam to the steam cylinders as in the original cycle. Also, the exhaust from the steam engine drove a turbo-blower to provide scavenge air. The *Eurybates* was converted to straight diesel drive in 1948 and after conversion was found to be more economical.

The original concept of a marine diesel engine was the large, slow-running direct-drive unit of between 90 to 160 rpm, but in the early 1920's two alternative drives made their appearance: the medium-speed engine of between 400 to 700 rpm coupled in pairs through fluid, air or electro-magnetic slip couplings into a reverse-reduction gearbox on a single propeller shaft, or a set of reasonably high-speed engines driving alternators and providing current to drive a screw-shaft motor. In either case, the total horsepower was broken down into a number of small independent trunk-piston engines, any one of which could be stopped for repair or overhaul, or simply for economy, without impairing the ship's navigability.

Although the diesel-electric system was developed in the 1920's, it was first in use as long ago as 1913 in the *Tynemount*, built in the north-east coast area, but in service on the Great Lakes. This ship had two sets of 6-cylinder air-injection engines running at 350 rpm, driving alternators with overhung exciters, and a single-screw motor. Unfortunately, this machinery was not successful and it was removed and replaced by a conventional steam engine in 1914. Incidentally, as the diesel engine became established in the 1920's and 1930's, more and more existing steamers were converted to motor, but we must not overlook the fact that engine conversion was not always in the same direction. There were many instances of motor ships being converted to steam reciprocating or turbine drive, two of the largest and best known being the *Asturias* and the *Alcantara*, both large passenger steamers.

Another concept of the oil-engine drive was the paddle engine, considered to be a fairly modern development, but in fact almost as old as the marine oil engine itself. It was in use on German rivers as far back as 1912: a small high-speed engine set athwartships driving the paddle shaft through reduction gearing.

This chapter on marine diesel machinery has grown out of all proportion to the limits originally intended for it, in spite of keeping descriptions of the more interesting cases to an absolute minimum. The field is so vast and there are so many different types, shapes and styles of oil engine that it is quite impossible to do justice to the subject in a single chapter or, for that matter, in a single book. But before closing the chapter I must give a brief outline sketch of the (SS or MV?) *Holzapfel I*, the first and possibly the only gas-engined vessel in the merchant service. Now the gas engine had been used as a power generator in land plant for many years—it was, in fact, the forerunner of the oil engine and the grandfather of the diesel engine. Plant up to 3500 BHP on a single crank were commonplace but so far no attempt had been made to introduce the system for marine propulsion. However, the Holzapfel power plant at Felling-on-Tyne had been working for years with a fuel consumption of 3/4 lb bituminous coal/IHP/hour, and the owners decided to build and power a coasting collier with similar type machinery. Accordingly, the *Holzapfel I* was built by J. T. Eltringham of South Shields in 1911 and engined by E. S. Hindley & Sons of Bourton in Dorset. She was 120 feet long, 22 feet wide and 11 feet 6 inches deep, with 350 tons displacement, her engines being fitted aft. The engine itself, was a six-cylinder unit working on the four-stroke cycle with side valves. Cylinders were $10\frac{3}{4}$ inches bore by 10 inches stroke, the speed being 460 rpm. Ignition was by spark plug and low-tension magneto, and the three forward cylinders were arranged for air starting, the compressed air being supplied by a small oil engine-driven compressor. A most unusual feature of this engine was its offset alignment on the désaxé principle.

As the engine was not reversible, a Föttinger power transformer was introduced into the line of shafting abaft the flywheel to provide for either ahead or astern running.

There were two gas-producer units, each of 100 BHP capacity, complete with dust boxes and coke scrubbers, and in service both were in use at the same time. To start up from cold, a steam jet was used on the producers until such time as the engine was run up to speed, it being possible to do this on stand-by with the Föttinger clutch in neutral position. To provide steam for this initial start and for the engine-room auxiliaries, deck machinery, etc., a small vertical donkey boiler was installed on the starboard side.

These two gas producers were housed in a small stokehold in the forward port corner of the engine room and each consisted of a small jacket boiler surrounding a coking furnace, with water cooler and coke scrubbers to cleanse the gas. In action, the engine drew a mixture of air and low-pressure steam through the incandescent fuel and the gas so formed passed into a water-bubble cooler, up through a wet-coke scrubber and into the gas-mixing valve on the engine. The steam was thus decomposed by the heat and its constituents hydrogen and oxygen were set free to mix with the carbon from the coal.

$$\text{Thus: } 2C + O_2 \rightarrow 2CO$$
$$\text{and } C + H_2O \rightarrow CO + H_2$$

In fact, several combustible gases were produced by this catalysis: carbon monoxide (CO), hydrogen and marsh gas (CH_4).

Coal consumption over several voyages worked out at about 25 cwt in twenty-four hours, equivalent to a figure of 0·75/IHP/hour, roughly half that of a conventional reciprocating steam engine of the same power.

11 The gas turbine

Logically, the gas turbine follows directly upon the diesel engine as naturally as the steam turbine succeeds the steam reciprocating engine, and for the same reasons.

Technically, the gas turbine is merely an extension of the hot-air machine, that caloric engine so beloved of early Victorian engineers and demonstrated in such dramatic fashion in the hot-air paddle ship *Ericsson*, where the crankshaft was driven by four pistons each 14 feet in diameter.

Historically, however, the gas turbine owes nothing to any of these, it was born in the eighteenth century long before either the internal-combustion engine or the steam turbine were even thought of. John Barber, in 1791, invented a patent engine using inflammable air for the purpose of procuring motion. Coal or wood gas from a retort and air were pumped into an accumulator kept at constant pressure by a head of water from an overhead tank, and upon being ignited 'issued forth with amazing velocity in a continuous stream against the vanes of a paddle wheel'. The air and the gas pumps were worked by beams off the shaft, and water injection was used for cooling. The inherent weakness of the gas turbine was apparent even in the early stages, as the gas and air compressors absorbed all the available energy in the turbine shaft and left none for useful work.

Barber's turbine was a genuine gas turbine just the same, where the power gas was generated within the combustion cycle: the gas generator (retort and combustion chamber) and the prime mover (turbine wheel) both formed part of the one machine unit, in contrast to the steam turbine which depends upon an outside supply of steam to make it work.

It is possible, of course, to classify gas turbines according to type, cycle, performance or manufacture, but the categories would tend to blur about the edges as there is no exact definition of a gas turbine. An extreme case is the turbo-expander which uses high-pressure natural gas and exhausts it into the gas main for domestic use, thus obtaining power at the expense of pressure, while the gas retains all its natural properties, a type of pass-out turbine, in fact. This reducer is a turbine running on gas pressure but it is not what one has in mind when describing the gas turbine.

In the Merchant Navy the gas turbine can be used in three distinct ways:

(a) as an exhaust-gas turbo-blower to supply air under pressure for scavenge or supercharge in main and auxiliary diesel engines.
(b) as a propulsion unit in association with two or more free-piston gas generators.
(c) as a self-contained unit supplying power to main or auxiliary alternators.

The exhaust-gas turbo-blower was introduced in the late 1920's initially as a supercharger on the large four-stroke main propulsion diesels and later as a scavenge pump for two-stroke engines. Its merit lay in using part of the heat from the exhaust gases, which normally went to waste up the funnel, to provide extra air to the cylinders, allowing more fuel to be burnt at constant pressure and thus enabling the engine to develop considerably more power without any increase in scantlings. To the ship-owner it was, in fact, something for nothing, and the turbo-blower has enabled the diesel engine to be up-rated to an extent undreamed of twenty years ago. Where waste-heat boilers were fitted in the exhaust cycle they could still produce sufficient steam at sea even after the exhaust gases had passed through the turbo-blower.

This turbo-blower, although built with watchmaker's precision, is an essentially simple machine. It consists of a spindle with a single-stage turbine wheel at one end and a centrifugal compressor wheel at the other end, the turbine wheel running in its separate water-cooled casing. The exhaust gases from the engine cylinders are piped to the turbine nozzle box in such a manner that pulsation is damped out, and as the gases are of a reasonable temperature, between 450–600°C according to type of engine, and at low pressure the turbine wheel requires no special cooling. The blades, in common with gas-turbine practice in general, are slotted into the wheel with a fir-tree dovetailed root, and they are usually shaped on the Vortex theory, having an impulse blade profile at the root and a reaction blade profile at the tip. They are also usually set into the wheel disc to allow of a small amount of tip movement to dampen out the stresses induced by the extremely high peripheral speed: some of the smaller blowers rotate at over 30,000 rpm, although admittedly in the very small units the turbine blades are integral with the wheel, the assembly being produced as an investment casting and similar in shape to the compressor wheel.

The gas turbine in association with free-piston gas generators, as a propulsion unit, is a newcomer to the marine engineering industry, being barely twenty years old, and it can be installed either as a reversible unit, as in the case of the steam turbine, or unidirectional, with a controllable pitch propeller, but in all cases with reduction gearing.

The propulsion machinery consists of two or more gasifiers delivering into a gas collector or receiver, thence through a proportioning valve into the turbine nozzle belt and exhausting into the atmosphere, with or without a waste-heat donkey boiler in the exhaust path. The gas generator or gasifier is constructed as a free-piston engine, basically a two-stroke uniflow scavenged, opposed-piston diesel engine supercharged by direct-connected compressor pistons. In this system all the reciprocating work produced in the diesel cylinder is absorbed by

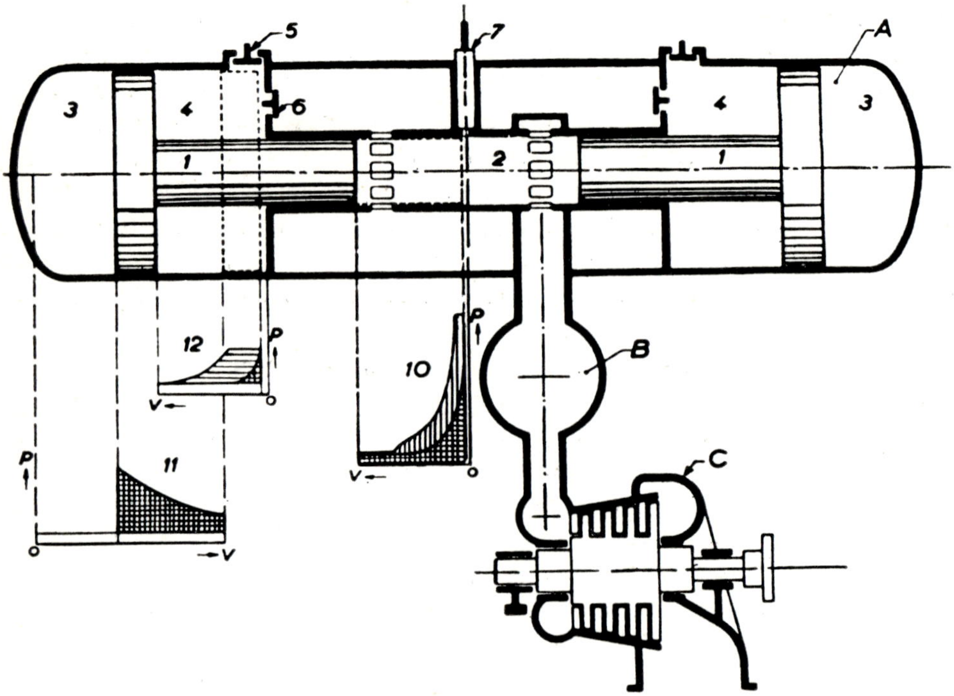

67 Diagrammatic sketch of a free-piston gas generator and turbine

the compressor pistons, while all the air from the compressor cylinders passes into the engine cylinder at pressures of up to 40 lb as scavenge air to be mixed with the fuel. The compression pressure is in the region of 1000 lb with a maximum combustion pressure of 1700 lb and with an early exhaust-port opening the exhaust pressure enters the receiver at a pressure of up to 50 lb. In addition to supplying the air for combustion, the compressor pistons also store up energy in the cushion cylinders to stop the diesel pistons at the end of the outward (power) stroke and return them to the inner position on the compression stroke. The cushion cylinders also act as starting cylinders by barring the pistons into the outer position and admitting starting air into the cushion spaces.

This free-piston engine was invented by the Marquis de Pescara primarily to provide compressed air for a helicopter rotor, and his first compressor was introduced in 1925. In its original form, it was unsymmetrical, i.e. it had a cushion cylinder at one end with a compression cylinder at the other, the compression piston acting as a scavenge pump on the inward stroke and supplying compressed air for industrial purposes on the outward stroke, the weights of both sides of the engine being equal. The engine thus had the three prerequisites of the gasifier: combustion cylinder, compressor cylinder and cushion cylinder.

The pistons were linked together to a central trunnion or shaft to avoid an overstroke and danger of striking the cushion cylinder heads, and the oscillation of this central shaft with its cams provided the motion for the fuel jerk pump. To start this compressor, the pistons were wound outwards through their linkage against a strong spring and then released to form the initial compression.

The present-day gasifier differs little from the original in basic design except that cushion cylinders are fitted at both ends of the machine and that the end product is not compressed air but exhaust gas at high pressure and temperature. The diesel piston is direct-connected to the compressor piston: the high combustion pressure acting on the engine pistons overcomes the inertia of the moving parts but does not require extra strengthening in the engine framework, while the hoop stress induced in the combustion space by the radial pressure loading is taken care of in the cylinder design.

The cushion cylinder is under pressure from the entrapped air at all times and the pressure fluctuates between about 35 lb on the out stroke to 95 lb on the cushion stroke. The power output of the gasifier depends on the length of stroke and this in turn is controlled by the amount of fuel injected by the four or six injectors placed around the combustion belt, taking fuel from a single pump driven by the engine linkage shaft. This linkage is merely a device to retain the two pistons within their strokes relative to the cylinder mid-point.

The power output of the gasifier is proportional to the mass flow m of the gas, its temperature T and the adiabatic heat drop, in accordance with the formula:

$$\text{Gas HP} \propto m \times T \left[1 - \left(\frac{P_2}{P_1} \right)^{\frac{r-1}{r}} \right]$$

and the combined effect of these three variables means that a very small change in stroke can cause wide divergencies in the power produced. Thus, by reducing the length of stroke by about 20% a drop of 75% from full load can result.

Unfortunately, it is impossible to have nil output in a Pescara gasifier without stalling. There is a minimum effective stroke beyond which the diesel cycle cannot be maintained, i.e. the practical necessity for the piston to uncover the exhaust ports to obtain good scavenging, and this results in a considerable quantity of hot exhaust gases being supplied to the collector and thence to the turbine *even when the latter is stopped.*

To cope with this awkward supply of unwanted gas, the proportioning valve

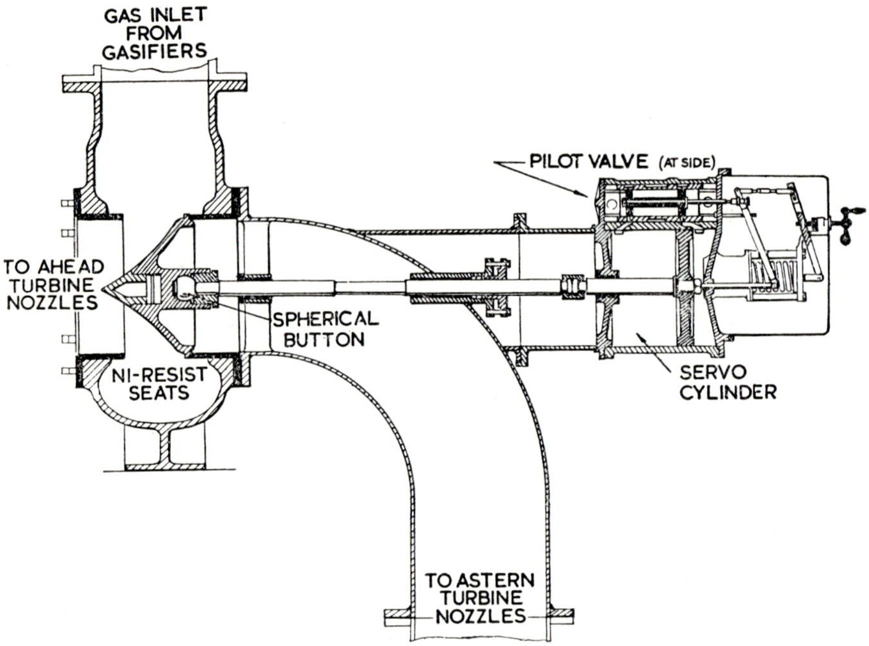

68 Gas-proportioning valve

is used as a manœuvring valve between collector and turbine. This is a double-ported valve chest, where gas from the collector enters between the seats and proceeds to the ahead or the astern turbine nozzle boxes depending upon the position of the valve lid, controlled through a servo-motor from the control platform. When manœuvring or on standby, with the turbine at rest and the gasifiers at minimum stroke, the valve lid is in mid-position and the gas is distributed to both ahead and astern turbines to cancel out opposing torques.

The turbine itself is a lightly built unit incorporating the ahead and astern rotors on the same shaft, the ahead rotor having up to six stages, the astern rotor being either single- or two-stage velocity-compounded. The turbine blading is either supported on stage wheels or carried on a gashed rotor, and is generally of the reaction type.

The fuel consumption for the Pescara-turbine cycle is fairly good, although it has not yet been able to equal that of the direct-acting diesel engine. However, it has been installed in a fair number of ships, either as built or as a conversion, on many different trades, and has proved quite successful: ships ranging from 10,000 tons deadweight to quite small inshore vessels.

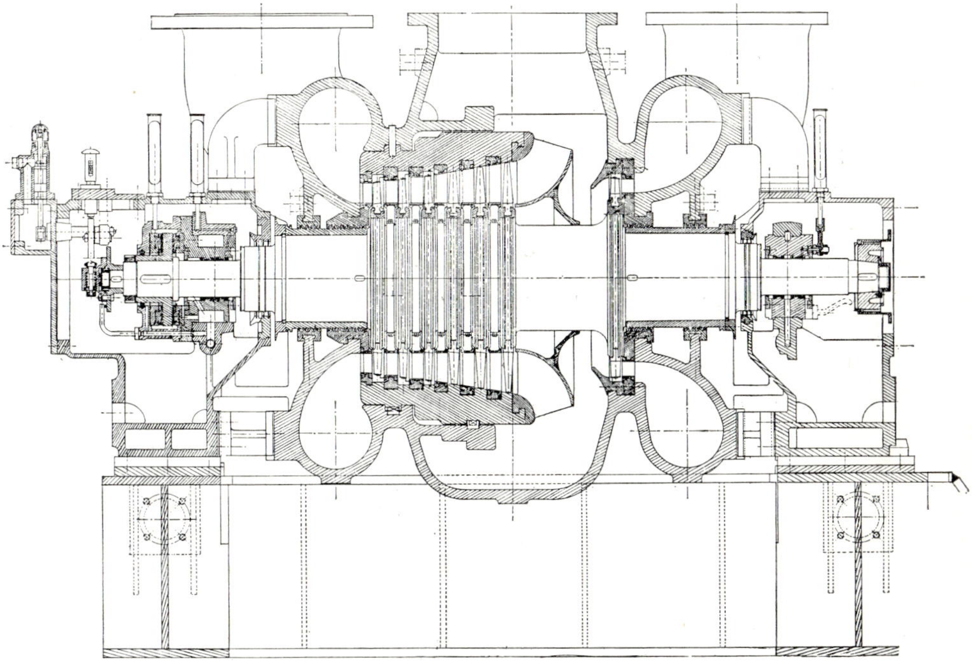

69 Cross section of reversible turbine driven by gas generators

This type of gas turbine has also been used in association with direct-drive diesel engines, as in the case of the twin-screw passenger vessel *Fritz Heckert*. Here, each shaft is driven by a 2300 BHP eight-cylinder two-stroke diesel engine with an electro-magnetic slip coupling, and by a reversing gas turbine driving the propeller shaft through double-reduction gearing, there being a fluid coupling interposed between turbine and primary-gear shaft. Each turbine develops 2800 BHP at 5500 rpm and is supplied by three gasifiers.

Finally, while acknowledging the simplicity of the gasifier turbine installation, one might be tempted to point out certain disadvantages in the system as they affect the operator. Any propulsion unit working in a confined engine room with a gas at 500°C, however well lagged, is bound to be unduly hot and uncomfortable. The double-reduction gearing between turbine and propeller shafting, however well cut, is invariably noisy and its high-pitched scream has an unpleasant effect on the nervous system. This is not helped out by the inevitable noise from the free-piston gasifiers, but the most unpleasant feature of all is the physiological effect of the air pulsation on the human frame, where the high-speed thumping of the compressors on the ears is accompanied by an equivalent pressure beat on the chest which tends to inhibit quick thinking.

So far we have discussed gas turbines that depend upon an independent source for their power gas, in one case a diesel engine, in the other case a diesel-type compressor unit, and in theory at least, the turbine systems just discussed were not very different from the steam turbine with its concomitant boiler as a source of energy. The real gas turbine, as now understood, is a self-contained unit with its own compressor, combustion chamber and power turbine with gearbox normally fitted on to the one base and in that respect it resembles the internal-combustion engine in that it is undependent of any other working equipment. As stated at the beginning of this chapter, it is merely a refined form of the hot-air engine and like any other heat engine, it must obey Carnot's principle in that the amount of heat converted into work depends solely upon the working range of temperature.

Many inventors in the past have tried to make a workable hot-air engine, among whom Stirling, who patented an engine to run on the constant-temperature cycle, and Ericsson, whose engine ran at constant pressure with a regenerator or heat exchanger between engine and compressor cylinders. Joule's system showed most promise, although it was never developed for use. It consisted of a power piston and a compressor piston in tandem, an externally fired heater and a water-cooled refrigerator, all situated in a closed-cycle air path. The cycle was quite simple: the air in the heater at temperature T_1 was admitted to the power cylinder and expanded adiabatically to the pressure obtaining in the cooler. At the same time cold air at T_2 from the cooler was drawn into the compressor and compressed adiabatically until its pressure reached that in the heater, where its temperature was raised to T_1 again. With heater and cooler of fairly large volume this cycle ran at a reasonably constant pressure and this is a feature of the gas turbine which is sometimes difficult for the steam engineer to appreciate. In the case of steam where temperature and pressure are interdependent the highest pressures and temperatures are at the beginning of the cycle, both dropping in value as the steam progresses through the engine towards the condenser. In the gas turbine, the highest pressure and temperature of the gas are halfway through the cycle, the compressor providing air to drive the turbine to rotate the compressor to provide air to drive the turbine, etc., and at first blush it does appear to resemble perpetual motion.

In the original hot-air engine, the mechanical efficiency was very low and a reasonable fuel consumption, in 1845, worked out at 2·7 lb/BHP/hour. In a poor engine the compressor took most of the power developed by the working cylinder, leaving a bare margin for useful output. Designers were up against the same problem that worried the early diesel engineers faced with a blast-air compressor that absorbed up to 10% of the engine power. But with the early gas turbines, the compressor required up to 75% of the power. Although there would appear to be no practical solution to this parasitic load, at least we have now been able to reduce it to little more than 50% of total power.

The earliest attempt at a practical self-contained gas turbine was made by Holzwarth in 1908 with the help of Körting Brothers and Brown Boveri Co. This was a genuine internal-combustion turbine in that it generated its own gas in a series of bombs or combustion chambers discharging directly into the nozzle box, and it was unique in that it ran on the constant-volume cycle.

The turbine, of about 1000 HP, was set vertically in its frame with a dynamo mounted above it. The spindle carried a two-stage single-wheel impulse turbine in a casing with ten nozzles placed around its perimeter, each nozzle being supplied by the gases from separate combustion chambers. Each of the ten combustion chambers was pear-shaped, narrow end towards the nozzle, and contained three valves: for air, gas and outlet to the nozzle. These thirty valves were all controlled by hydraulic servo-motor system, while firing was by means of HT ignition. As a safety measure, the outlet valve from each bomb to its nozzle, though controlled by a timing device could also act as a non-return relief device.

The turbine casing and the combustion chambers were all water-cooled at high temperatures, and the steam generated from the system was used by a separate turbine to drive a scavenge blower. The installation was thus completely self-contained, and it ran at 3000 rpm with a maximum firing pressure of 120 lb. Although attempts were made to develop this type of turbine, it does not seem to have made the grade and attention has been switched to the constant-pressure turbine. Although this type had been known and experimented on for many years, the first really successful installation was set up in Switzerland in 1940, and it is fair to add that this success was due not so much to the turbine as to the design of air compressor.

Sir Charles Parsons devoted much of his spare time experimenting with axial air compressors and his first commercial turbo-compressor was made in 1900 and ran until 1916. In 1938 the Parsons Marine Steam Turbine Co. built an experimental gas turbine that developed 500 BHP, at 6000 rpm on Pool fuel, with a working pressure of 50 lb. It had a heat exchanger wherein the compressor delivery air was heated by the exhaust from the power turbine, and the turbine rotor was a solid forging with eight rows of blades while the axial compressor had twenty-five rows of blades.

The combustion chamber, one of the most important features of the turbine, was also developed at this time, with part of the compressed air being diverted outside the flame shield or liner to cool the chamber walls.

The closed-cycle turbine, as developed in the Escher Wyss system, is a pure hot-air machine, the air being pre-cooled before entering the compressor and pre-heated in an exhaust-gas heat exchanger after compression. It is then heated in an oil- or gas-fired combustion chamber before entering the power-turbine nozzle box. In this turbine cycle the air circulates round the system without mixing with the combustion gases, and there is consequently no deposit on the blading or in the heat exchangers. Also, as the air is held at pressures well above average, the components can be kept to a reasonable size. The disadvantage, however, at least in a cargo vessel where the engine room must be kept as small as possible, is in the size of the heat exchanger and pre-cooler.

The modern gas turbine seems to have settled for an open-cycle constant-pressure type without heat exchanger: the latter is sometimes a mixed blessing, in view of its bulk. But apart from space problems, while a heat exchanger in the cycle can increase the efficiency from, say, 18 to 22%, the effective horse power can be reduced to about 75% of full load. Thus, to obtain the specification power, a larger-size turbine would need to be fitted if heat exchangers are contemplated.

Small gas turbines suitable for auxiliary-power generation on board ship could be produced as a single-turbine unit where a two- or three-stage wheel would drive both the compressor and the output shaft. This self-contained unit can be made the more compact by using a single-stage centrifugal compressor with inducer, and the relatively small bedplate required for prime mover and alternator should appeal to ship-owners.

Turbines developing higher powers are usually split into two sections, both mounted on the same baseplate. One half, the generator section, carries the HP turbine direct-coupled to an axial compressor, while the LP turbine is the business end of the machine, it delivers the power to the output shaft through reduction gearing, usually of the orbital or epicyclic type. However, there would appear to be no hard and fast ruling on single or double rotors when driving machinery at constant speed, merely a convenience in design. Also, although axial compressors, being more efficient, are the rule for large turbines, single- or two-stage centrifugal compressors can be used where space is really important.

The operating cycle of an industrial gas turbine is refreshingly simple in theory. Air at ambient temperature and pressure is drawn into the compressor and delivered to the combustion chamber at, say, 200°C and 50 lb pressure,

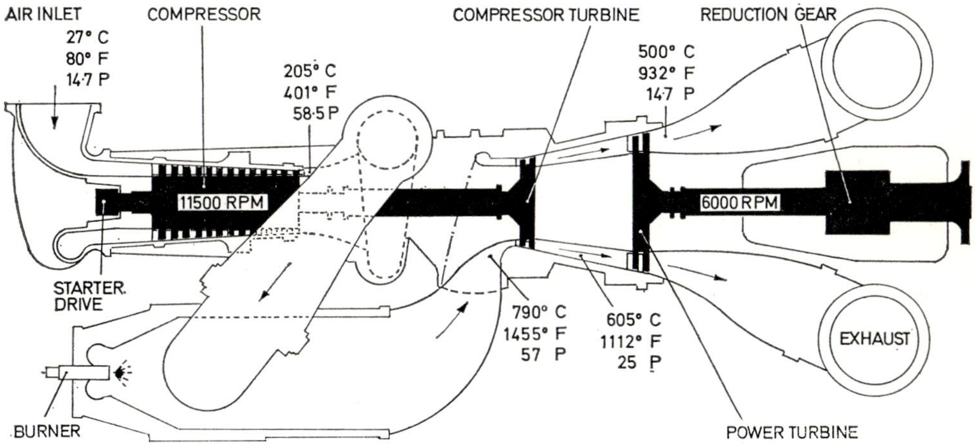

70 Section through 1500 BHP gas-turbine showing typical pressures and temperatures throughout cycle

there to mix with the injected fuel at constant pressure but with its temperature now increased to about 800°C. This mixture of combustion gas and excess air then passes to the HP turbine, usually a two-stage wheel driving the compressor, then to the LP turbine coupled through orbital gearing to the output shaft, finally exhausting to atmosphere at a fairly high temperature, about 400°C.

Now the sole aim of the turbine designer is to translate this simple theoretical cycle into useful power; at all costs he must prevent his machine from degenerating, like Cayley's, Stirling's and Ericsson's hot-air engines, into a mass of clanking and grunting ironmongery that absorbs all of its internal energy to drive its own air pump, with a mechanical efficiency of zero. Obviously a reciprocating or rotary-vane compressor could not be used, their frictional losses are too great and they could not be run at turbine speeds and the only really efficient high-speed compressors are the axial and the centrifugal. Of the two, the axial machine is the more efficient although it takes considerably more space, and it consists of a cylindrical rotor equipped with a number of rows of blades: ten to fourteen, each row of blades rotating between a row of fixed blades in the stator, the blades decreasing in size towards the pressure end. It is in some respects similar to a steam reaction turbine in reverse, and this likeness is enhanced by the end row of stator blades fitted in such a manner that they straighten out the rotating flow path on its way to the combustion chamber. As the air travels through the various stages of the compressor, rotating at up to 10,000 rpm, its velocity is converted into pressure by the stationary blades and is further increased by the diffuser at the end, and the air enters the combustion chamber at the maximum cycle pressure, about 50 lb.

The fuel injection is continuous, as in a blowlamp, but not all the air delivered by the compressor is required for combustion; a considerable proportion by-passes the flame to cool the combustion chamber walls, while another portion is tapped off from the compressor HP stages to cool the turbine blading, running at some 6000 rpm.

The turbines themselves are of the accepted impulse-root, reaction-tip section set into their discs in a fir-tree dovetail, and are given a certain amount of tip rock, as in the case of the exhaust-turbo-blower wheels. This tip rock, up to $\frac{1}{32}$ inch, manifests itself in a most unusual way when the turbine is shut off and is coming to rest: the blades emit a soft ticking sound easily recognised by the gas-turbine engineer but most disturbing to a steam engineer, whose turbine blades are rigidly fixed and are expected to remain silent and stay put at all times.

The present-day small gas turbine is not outstanding in fuel consumption, mainly because of the terrible parasitic load from the compressor, and an average unit will be lucky to run at 0·70 lb/BHP/hr, rather high by comparison with modern steam practice and riotously extravagant when compared with the diesel engine. It is, however, extremely economical on lubricating oil compared with a diesel engine and this point must never be forgotten when lauding the low fuel consumption in a motor vessel. Some twin-screw vessels can consume up to 400 gallons of lubricating oil a day, and this cost, say, £40,000 per annum, must be added to the fuel bill. It is also very simple to operate, requiring little ancillary equipment, just the independent oil fuel and lubricating oil pumps and a source of power for the starting motor, usually electric, and itself of modest dimensions. The starting operations are so simple that the gas turbine is, of all prime movers, that best suited for automatic or remote control: starting up fuel and lubricating pumps, then starter motor, running up to predetermined speed, cutting in fuel and complete shutdown are all unsophisticated operations suitable for remote controls.

This turbine also has catholic tastes in fuel: it can be run on oil fuel, town gas, or, at a pinch, on peat that has been put through an atritor grinder and reduced to a fine dust. Moreover, it has one cardinal virtue which places it above all other heat engines for industrial use, especially in the tropics: it requires no cooling water. Possibly for this very reason, the gas turbine has made very little impact in the field of merchant-ship propulsion, where the supply of cooling liquid has never been at issue, and its application in transport has hitherto been confined to the aeronautical industry and, latterly, naval vessels, in both cases because of its astounding power/weight ratio.

The Navy experimented with the turbine as long ago as 1946 with seemingly satisfactory results. In one vessel a triple-screw patrol boat propelled by three internal-combustion engines of 1300 HP each, the centre engine was removed and replaced by a gas turbine, although it is suspected that the new prime mover occupied more space than the original motor. Since then, the Navy have developed this turbine until it is now a recognised feature of many high-speed naval vessels. However, in a fighting vessel, the problems of reliability, weight and economy cannot be considered in the same light as for a merchant vessel. Here, weight is not that important, but economy is—very important indeed. The cargo ship's function is to transport goods to the profit of the shareholders, not to promote a new form of propulsion. Consequently, merchant ship-owners have tended to drag their feet in adopting this new prime mover although as we have seen in previous chapters it was they, rather than the Royal Navy, who financed the various developments throughout the long history of marine engineering.

Probably the earliest adoption of marine gas turbines for auxiliary power was in a number of Blue Funnel ships in 1956 when they were fitted out with 350 kW emergency sets as an addition to their normal electricity generating capacity. These were driven by two-stage open-cycle turbines running at 15,000 rpm geared down to 1500 rpm at the output shaft, and with a single-stage centrifugal compressor, they formed a fairly compact unit. If I remember correctly, they operated on push-button controls from outside the engine casing. They were not the first gas turbines to be fitted in a merchant vessel, however; that honour belongs to the Shell tanker *Auris*, a 12,000–ton vessel built in 1948. Her original machinery consisted of four 1100 HP diesel engines driving 850 kW alternators supplying current at 1600 volts to a 3750 HP synchronous motor on a single propeller shaft. In 1951 the starboard inner diesel engine was removed and replaced by a 1200 HP open-cycle gas turbine with a view to ascertaining the suitability of this type of drive for ship propulsion.

The turbine installation was of the two-spindle regenerative type, the HP turbine driving the compressor being placed above the LP turbine direct-coupled to a 950 kW alternator feeding into the main propulsion motor. The HP turbine was started by an electric motor on the free end of the axial compressor spindle and the gas-flow path led from the compressor through an exhaust-heated tubular heat exchanger. Temperatures and pressures at the HP turbine inlet were 600°C and 50–60 lb, dropping to 450°C at the LP or power-turbine inlet.

The *Auris* was completed and back in service by October 1951, and in March 1952 she crossed the Atlantic with the gas turbine only feeding the main propulsion motor, the turbine itself running smoothly on boiler oil, and keeping up a ship speed of $7\frac{1}{4}$ knots. In fact, we have it on reliable authority that this turbine was more dependable than the piston diesels and on more than one occasion it was responsible for the safe return to port.

After a year's experimenting with the gas-turbine unit, when it had run for more than 20,000 hours of which nearly 7000 on residual fuel, the owners decided to try the turbine out in a full-scale commercial venture, to be competitive with the steam turbine-driven ships of the same class in their fleet. They therefore placed an order for an 18,000-ton twin-screw tanker, the *Hemisinus*, to be propelled by two sets of gas-turbine engines on the open cycle, and work was put in hand to design this machinery on the basis of data and experience acquired in the *Auris* experiment.

By 1955, however, a train or circumstances, partly economical and partly technical, forced the owners to cancel the project for the *Hemisinus* and transfer their attention to the *Auris* which by now required some new machinery. The *Hemisinus* was accordingly equipped with conventional steam turbines and her port gas-turbine installation, already well advanced, was destined for the *Auris*. This vessel was accordingly withdrawn from service, and her main propelling machinery entirely removed. The shafting only was retained fitted with a propeller of the original diameter but with coarser pitch and larger developed area, to take account of the greater horsepower, given as 5000.

This installation was designed as an open-cycle two-shaft unit, the HP turbine driving an HP compressor and the LP turbine driving both the LP compressor and the screw shaft through gearing. The gas path was simple. Air was drawn through a spray chamber into the LP compressor, thence through an intercooler to the HP compressor and on to a heat exchanger where it absorbed a considerable amount of waste heat from the exhaust gases. From there it passed to the combustion chamber where it acquired the necessary heat to expand in series through the HP and LP turbines. The exhaust gas from the latter passed up through the vertical heat exchanger and into a waste-heat donkey boiler before escaping to atmosphere.

The HP and LP turbines were of five stages and eight stages respectively, while their compressors carried sixteen and twelve rows of blades. The combustion chamber was refractory lined and was placed vertically below the HP shaft, the object being to prevent any broken refractory from being entrained with the air flow.

The power turbine, i.e. the LP turbine-compressor unit, drove the main shaft through 32:1 double-reduction gearing, there being three hydraulic couplings on the primary shaft to drive ahead, astern and a take-off shaft driving the 550 kW alternator to provide power for the cargo pumps. The quill shaft through the primary pinion had a hydraulic coupling at each end, and ahead or astern rotation depended upon the appropriate coupling being filled. In the case of the astern coupling, the oil from the driving half was deflected through a row of stationary vanes to reverse its tangential flow and to rotate the driven half in the opposite direction. These hydraulic couplings were not positive drives: the astern unit had an efficiency of 60–65% and was considered suitable for its purpose. The ahead coupling had a small slip, but as this nevertheless represented a power loss, an oil-operated friction clutch could be engaged on 'Full Away' to give a positive, straight-through drive.

A steam starting turbine engaged the HP turbine rotor shaft at its forward end through a slip clutch which enabled the installation to be re-engaged for restarting without having to wait until rotation had died down to a complete stop.

At full power, given as 4400 SHP, with propeller turning at 119 rpm, the HP and LP turbines ran at 5640 rpm and 3800 rpm, with inlet temperature about 620 °C and 520°C respectively, while the specific fuel consumption for propulsion only, based on mean propeller SHP, worked out at 0·556 lb/hour.

On trials in 1959 and in service the installation showed up a number of teething troubles although it is fair to say that most of these could be blamed on the ancillary equipment, seldom on the turbines themselves. In fact, although new propelling machinery had been installed, many of the old auxiliaries were retained and incorporated into the gas-turbine system on a sort of compromise basis.

The biggest problem to overcome was that of manœuvring, and it will be recalled that the LP turbine drove the LP compressor as well as the main shaft. Thus, when the propeller shaft was running at slow speed, the LP compressor was also running at slow revolutions. Now the axial flow compressor, more than any other type of rotary machine, is essentially a constant-speed compressor: its blading is designed for one air speed and one air pressure only, and any change in the precarious balance of revolutions, air mass and air pressure will most certainly bring about a drastic drop in efficiency. The compressor speed is decided by the turbine speed which in turn is dependent upon the amount and pressure of the air produced by the turbine—a vicious circle indeed. Also, the compressor blading is so designed that the stator vanes guide the high-speed air on to the rotor blades in a smooth axial flow when running at designed speed. Any sudden change in speed will upset this smooth flow of air through the compressor, the air will be deflected from the stator vanes at the wrong angle and some stages will choke and stall. The compressor has reached the surge point where efficiency falls to zero and where the rotor can shuttle in its bearings and shed its blades. The obvious solution is to fit relief or blow-off valves somewhere in the cycle to release the build up of air as it starts, and this can be either at the LP end of the compressor rotor, where the air can be released before too much work has been put into it, or at the delivery end, but in most cases, after surging, the turbine must be allowed to coast to rest and be restarted.

To safeguard the *Auris* turbines, blow-off valves were fitted to the HP compressor to release the surplus air at reduced revolutions and when going at dead-slow propeller speed, both ahead and astern hydraulic couplings were engaged, a manœuvre reminiscent of the Pescara-turbine cycle and its proportioning valve.

Cleaning the compressor blading was by slowing down and feeding ground coconut shells into the air stream, either dry or with fresh-water spray. The modern method is to use a mixture of paraffin and fresh water, although it must always be borne in mind that the mixture thus injected is a form of unmetered fuel to the driving turbine.

One of the surprising side-effects of this gas-turbine installation was the amount of fresh water collected in the intercooler; when the 2–inch-bore drain valves were opened, the accumulated water in the cooler often ran out of the discharge pipes at full bore for several hours.

The *Auris* was withdrawn from service in 1960 mainly because of her size—the utility tanker had now become the 65,000 tons dw and the smaller units, even the economical diesel-driven tankers, were no longer suitable for their purpose.

An even more ambitious experiment was undertaken by the US Maritime Administration in 1956 on four standard Liberty ships. These wartime-built ships were driven by steam reciprocating engines of 2500 IHP at a steady 10 knots; good dependable vessels in their day, which eventually became the delivery vans for most of the Western world. However, 10 knots maximum speed was of no use at all in the 1950's, and more and more of them were laid up awaiting a lucky break: another war possibly, as an alternative to the breakers' yard.

The experiment was intended to investigate the economics of increasing the Liberty ship's speed to 15 knots, i.e. comparable with the latest trend in shipping, by fitting engines of considerably higher power, and each of the four ships was to be equipped with a different type of prime mover. The *Benjamin Chew* was fitted out with geared steam turbines, the *Thomas Nelson* with geared diesels, the *William Patterson* with free-piston generators in association with gas turbines and the *John Sergeant* with true gas turbines.

The *John Sergeant*'s engine was a 6000 SHP two-shaft regenerative gas-turbine installation with both HP and LP turbines in the same casing. The HP turbine drove the fourteen-stage axial compressor at 6900 rpm while the LP turbine drove the variable-pitch propeller through double-reduction gearing. It is interesting to note that both HP and LP turbine were single-stage wheels, and that the fuel consumption at 16·8 knots is given as 0·51 lb/SHP/hr for all purposes on diesel oil. Starting was by steam turbine, and being an experimental ship, she had a 750 SHP emergency steam turbine available as a get-you-home unit.

Although the *John Sergeant* is still in being, it would appear that she has been withdrawn from service.

To the best of my belief, the first merchant vessel to be designed for and propelled by gas-turbine machinery as a paying venture and not just as an experiment, is the *Adm. Wm. M. Callaghan*, built in 1967. This twin-screw cargo ship was specially built in USA to carry military vehicles, and is a 7000-tons dw ship propelled at about 26 knots by turbine machinery of some 40,000 SHP. The two gas generators are Pratt and Whitney aircraft gas turbines adapted for marine service, and each generator provides the power gas to drive a free turbine connected to the propeller shaft through reduction gearing, the engine-room layout being so arranged that the generators can easily be removed ashore for overhaul.

The double-reduction gearing incorporates pneumatic reversing clutches in a most interesting manner. The turbine pinion meshes with both ahead and astern primary wheels whose quillshafts drive their respective ahead and astern clutches. These clutches in turn drive secondary pinions meshing with the main wheel and rotation of the propeller depends simply upon the clutch engaged.

Fuel consumption is stated to be equal to that of a modern steam-turbine installation, and assuming that the owners are agreeable to repair by replacement it would appear that the marine gas turbine is here to stay. Indeed the system used in the *Adm. Wm. M. Callaghan* has proved successful to the extent that it is now being adopted for use in several large fast container ships. Previous marine installations highlighted the reliability of the turbine itself—when things went wrong the trouble was almost always invariably traced to ancillary equipment. And, a point to be seriously considered is that although sea air may not be the best ambient for this sort of machinery, quite a number of industrial gas turbines have completed 100,000 hours of steady running without major overhaul, equivalent to seventeen years' trouble-free service in an average steamer.

12 Hydrofoils and Hovercraft

Hydrofoils and hovercraft are newcomers to the field of merchant shipping and they both differ widely from the conventional ship, and from each other, in their methods of propulsion and general behaviour when under way. Their characteristics and their type of propelling machinery are in general borrowed from the aeronautical industry rather than from orthodox marine-engineering practice, but as they will probably still depend upon dockyards for their maintenance, and as they are acknowledged by the various governing bodies and classification societies to be part of the Merchant Navy, they are included in this history to complete the overall picture.

Hydrofoils

This type of vessel stems from the hard-chine semi-displacement craft, where at planing speeds the total weight of the hull is supported partly by buoyancy, partly by the water pressure on the underside of the hull, i.e. the lift is partly static and partly dynamic, the ratio of one to the other depending upon speed for any given shape. The planing vessel usually has a flat bottom with a sharp-angled bilge to throw the wake clear of the ship's sides, and only sufficient deadrise to give her directional stability and prevent skidding. The forefoot, sides and transom are, or should be, clear of the water to avoid skin friction, and the buoyancy is provided solely by the bottom planking. In a planing craft we thus have the anomaly of buoyancy, which is a function of volume, being supplied purely by a surface with no depth. This buoyancy is given by the wedge of water displaced by the vessel's bottom planking at planing speeds, when the forefoot rises clear of the surface and the stern squats well down. The dynamic lift will be understood by reference to the age-old seaside pastime of throwing flat stones into the water at a low angle of incidence to make them skip over the waves.

The hydrofoil is and behaves like a conventional ship whilst at rest or when manœuvring in harbour, but being equipped with underwater wings or foils, she rises upon these when travelling at speed until the hull is completely clear of the water and is then supported only by the dynamic lift of the foils. In this case, buoyancy can conveniently be ignored. Hence the relationship between the hydrofoil and the conventional ship is comparable to that of the kite and the captive balloon—they both float in the air on the end of a string but for entirely different reasons.

However, the economics of hydrofoil operation are not comparable with the running of orthodox merchant ships, as both first cost and maintenance are higher

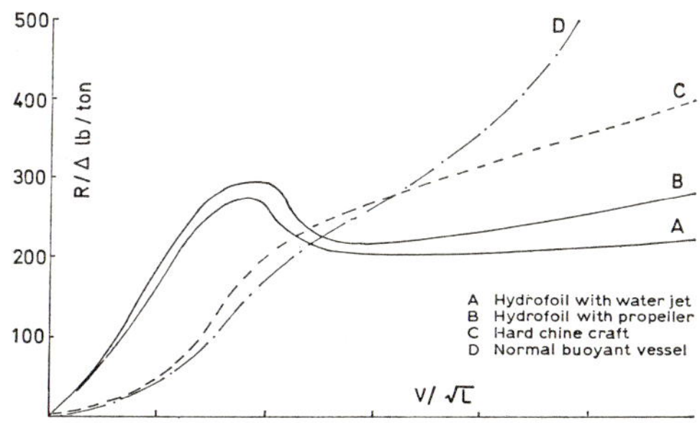

71 Resistance/displacement ratio curves for various types of small high-speed vessels

on a price-per-ton payload basis. The main characteristics of the hydrofoil are lightness and speed—very high speed—and these are viable only on special routes for special cargoes or passengers. Referring to the graph showing resistance/displacement ratios for various types of small vessels, it will be noted that the hydrofoil craft is competitive only at speeds above $4\sqrt{L}$. In practice, this means speeds of 30 knots and over, and this in turn determines the length or duration of the passage and the particular trade.

Compared with the conventional power-driven vessel, the hydrofoil has a very short history and its development has been influenced by, and kept in step with, that of the aeroplane. By fitting a boat with underwater wings and a powerful engine, the designer has succeeded in lifting the hull clear of the water at speed and thus very considerably reducing the skin friction which absorbs so much of the engine power. A higher speed can then be obtained from the engine and as the hull now rides clear of small waves, a much more comfortable trip ensues.

The earliest patent for a hydrofoil craft would appear to be that of the Comte de Lambert in 1891, when the inventor sailed a demonstration craft on the Seine to show how hydroplanes could increase a vessel's speed while using the same engine power. This vessel was driven by a steam engine and appeared to have four hydroplanes on each side. At speed these hydroplanes were lowered into the water and evidently lifted the hull partly clear of the surface.

Wilbur and Orville Wright, the pioneers of aviation, experimented with the hydrofoil idea in 1907 but dropped it to concentrate on the aeroplane.

The Italians were very much to the fore in the early days of hydrofoil development. Enrico Forlanini experimented for many years with ladder foils and in 1906 demonstrated a small unit on Lake Maggiore that reached a speed of 38 knots. These ladder foils, as built in later years, have their 'rungs' fitted at dihedral angles such that as the vessel rises above the surface, the lower part of one rung leaves the water as the top part of the one below starts to emerge. This is to ensure a smooth rising of the craft on her foils without any sudden change of lift and consequently without the associated bumpiness, and can be said to equate the weight of the hull to the draught through the various speeds.

Two other Italians, Guidoni and Crocco, experimented with light hulls supported on dihedral foils long before the outbreak of World War I, and it is interesting to note that the impetus in the study of hydrofoils was given by inventors associated with aeroplane design. At the end of this war, Graham Bell and his partners Casey, Baldwin and Philip Rhodes developed the Hydro-Drome series, and as early as 1919 one of these, powered by a couple of Liberty engines each of 350 HP, attained a speed of 61 knots with a total weight of 5 tons.

Several patents were taken out in Britain in the 1930's, notably by the John Samuel White Company in the Isle of Wight and by Grunberg, of France, the latter introducing the fully submerged main or support foil which is fixed to the body, with a pair of planing surface foils or floats to control the incidence. Also in that decade, Baron Schertel von Burtenbach introduced the hoop-type foil which at a later date developed into the Vee type in association with anhedral struts to improve stability.

The prime characteristic of the hydrofoil is speed with its corollaries of lightness and high engine power. Apart from naval and rescue craft, therefore, where economy is not the major factor, its application in the Merchant Navy would be restricted to the first-class passenger trade in relatively smooth waters, a high-class ferry service, in short, and large numbers have been built by the Russians for use on their large river network, some of them, in design study, intended for 100 knots or more.

There is no standardisation in the type, number or position of the foils although a general pattern seems to have developed whereby hydrofoils have either a combination of main foil amidships with a pair of surface skimming floats placed either forward or aft, or else two main foils, one forward, one aft. Again, the shape of foil varies from type to type, and can consist of a fully submerged foil, ladder foil, skimming foil or float and hoop foil.

With the craft at skimming speed there must be some means of ensuring 'flying height' and stability: directional, longitudinal and transverse, and although all these may be controlled manually in dead-calm waters it would be very difficult to steer a hydrofoil across rough or lumpy water without some type of automatic control. This is usually carried out by sensors combined with gyroscopes to operate hinged flaps on the foils, one on each side and one at the stern. As in

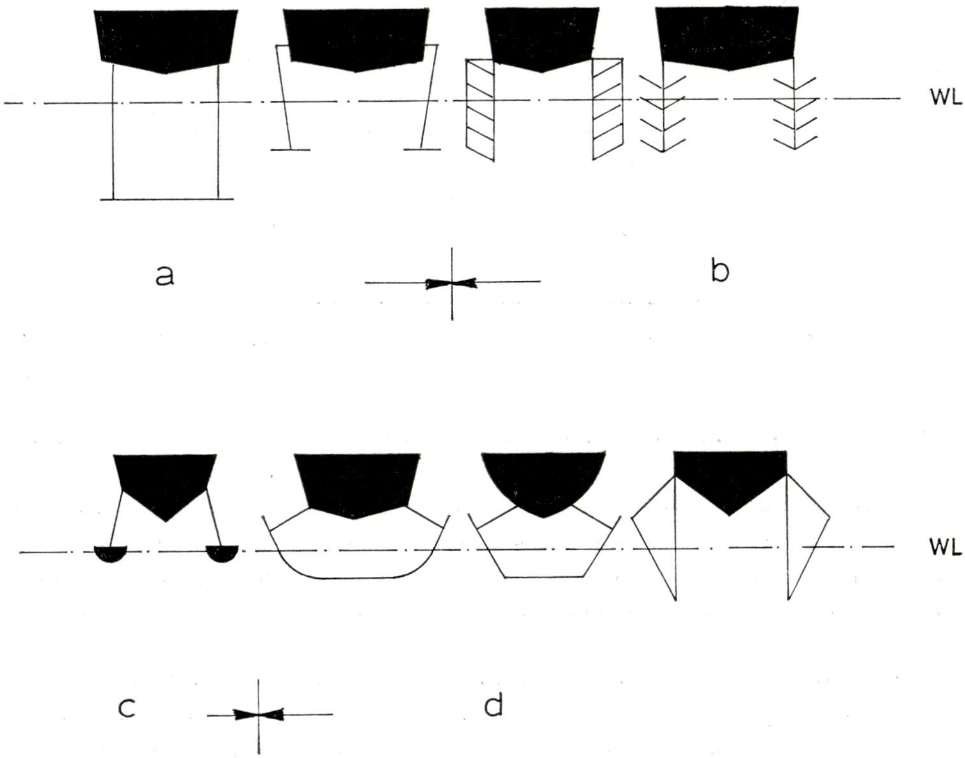

72 Various types of foils

the case of aircraft, the hydrofoil is most vulnerable in taking off, and any misalignment of the flaps could result in capsize; for this reason the system is usually set to fail safe and stop engines in the event of failure.

Steering a hydrofoil at speed is a tricky operation and it cannot be compared either with the aeroplane or the surface ship. In the first case, a large tail fin in a steady air stream maintains suitable directional stability and banking prevents side-slip; in the other case, the floating ship with its great lateral surface presented to the sea procures a tight turning circle. The hydrofoil cannot use either method and depends solely upon rudder for direction, and this has to be handled very carefully when making a U turn. The centrifugal force acts through the longitudinal centre of gravity, and if this is much abaft the main foil, the stern will tend to swing out of control. Possibly for this reason, some operators reduce power in turning, letting the craft sink into a buoyant vessel to turn in the conventional manner, then rise again upon the foils to resume the new direction.

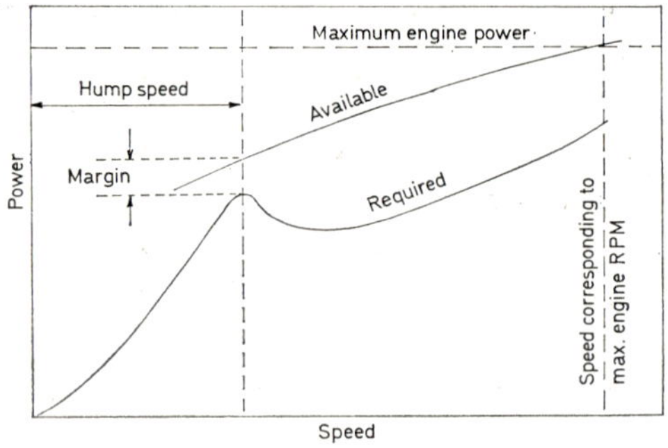

73 Power/speed curve showing margin between available power and required power at hump

As the rudder serves the craft both as buoyant and skimming vessel, it is made fairly long, to grip the water more deeply when manœuvring.

Hydrofoils can be propelled by one of several different types of prime mover, ranging from the light-weight diesel engine driving a propeller or water jet, through gas-turbine-driven air screws to the pure ram jet, and ranging in power from a few hundred BHP to 120,000 BHP.

In smaller craft, one diesel engine will suffice for harbour work and planing at speed, although in larger vessels, two or more engines may be required at full speed. Reference to the power/speed graph shows that the resistance increases rapidly for moderate increase in speed until the hull is foilborne, when the resistance drops dramatically then rises at a much slower rate for considerably high speeds. This hump is the designer's bogey as the configuration of the two curves: available engine power and required engine power, approach each other very closely at the hump. Any slight miscalculation of engine power or propeller characteristics and the vessel will refuse to rise from the water and remain a simple buoyant hull.

The choice of prime mover depends mainly upon the size and type of hydrofoil, diesel engines for the smaller craft, gas turbines for the larger. The diesels must of necessity be of the highly rated, high-revolution engines developing up to 2500 BHP continuous rating, weight being an all-important factor in the choice. Gas turbines, on the other hand, come into their own at the other end of the power spectrum, with individual powers of 10,000 HP, and with their unique power/weight ratio, they are the only solution to the problem of powering large hydrofoils. Unhappily, both highly rated diesels and gas turbines need servicing by replacement and this involves either a spare engine or rotors, or alternatively considerable delay in port until the engine has been serviced and refitted.

Unhappily, also, both these refined types of engine require a specialised maintenance crew working in expensively equipped workshops. The major components of these engines may be held together by high-tensile bolts and studs of the order of $\frac{3}{16}$ to $\frac{1}{4}$ inch diameter and their adjustment is effected by torque spanner, the joint tightness being gauged scientifically by micrometer on the bolt stretch. Manufacturers of highly rated diesels and gas turbines tend to frown on the time-honoured marine practice of hardening up slack crankcase nuts with a light flogging hammer at the end of each trip.

Apart from maintenance and first cost, however, the hydrofoil-owner faces a higher consumption in his engines, for lightness must be paid for. Diesel engines will run at between 0·38 to 0·42 lb/BHP/hour and simple-cycle gas turbines at anything between 0·55 and 0·65 lb/BHP/hour.

Propulsion can be by submerged screw, water jet or air screw, the latter being the more efficient at very high speeds. Although their large diameter makes them rather vulnerable, the air-screw propeller provides the simplest solution to the problem of high-speed travel over water.

With diesel engines, propulsion is mainly by propeller, either directly driven through a straightforward inclined shaft, or by right-angled (Z) drive; from the engine down through the foil strut, then horizontally out into the propeller, in principle not unlike an outboard-motor drive, and with the possibility of using the propeller to steer with.

It is the propeller, however, that gives the hydrofoil designer his greatest worry: cavitation, a phenomenon associated with practically all high-speed craft, but especially so in hydrofoils travelling at speeds of up to 50 knots, which would seem to be the upper limit for submerged propeller speeds. Cavitation can occur with propellers of almost any type of ship, it has been even noted in nine-knot tramps, but here it is due mainly to poor design and entrained air. It is in fact a characteristic of fast-revolving propellers, evident in an increase in slip and corresponding decrease in efficiency, where the propeller absorbs abnormal power after reaching a certain speed.

There are several ways of causing cavitation: by entry of air from the surface into the propeller, by dissolved air in the water being released upon arrival at the propeller blades or by a local fall of pressure on any part of the blade. The main cause, of course, is that as the blade velocity increases, the water pressure at its back decreases, roughly as the square of the speed, to less than the total head of water (as given by its draught) plus the atmospheric pressure, resulting in a partial vacuum at the back of the blade. As the pressure continues to fall with increasing speed, the water in the lower pressure area tends to evaporate leaving vapour bubbles or cavities at the blade surface. Upon contact with the propeller, these vapour bubbles burst but they have a very erosive action on the surface metal and eventually result in a honeycombed blade surface. In some hydrofoil craft it is usual to have a pair of propellers, one in use, the other being repaired, the change-over taking place at intervals of less than one month's service.

One way of combating cavitation would be to fit a controllable-pitch propeller, but this is seldom possible. In a straight drive, the propeller shaft would be too long to admit of an internal control rod without additional staying or suppprt, which would lead to added water resistance. In a right-angled drive there would appear to be no method of control through the several bevel gears.

The only way so far has been to design a supercavitating propeller using a special type of copper aluminium alloy with the blades wedge-shaped in cross-section. This blade form throws off the sheets of vapour and allows them to collapse harmlessly in the slip stream.

An alternative to the water propeller is hydraulic jet propulsion, where water is scooped into a duct from below the surface, directed into a rotary pump and then discharged in the form of a high-speed jet above the surface of the water. The scoop, which must take from well below the surface to avoid broken water, feeds into the high-speed pump, either axial or radial flow, which gives it considerable mechanical energy and discharges it through nozzles where the pressure energy is converted into kinetic energy in the form of a high-speed jet.

The gas turbine, with its very high power/weight ratio, is the obvious prime mover for the bigger hydrofoils, where its cost would be recouped in the larger pay load, but as already explained this ratio is attained only at the expense of a somewhat limited service life for some of its components. Duplication of prime movers is not the answer. Twin engines geared to the one propeller shaft are quite common in a merchant vessel, as failure of one engine still allows the vessel to proceed at nearly three-quarter speed. But in a hydrofoil failure of one engine would result in the power being insufficient to raise the hull on to the foils. To date the answer appears to lie in a compromise in the turbine arrangement: an aircraft turbo-jet engine to act as generator or gas producer, driving a marine type free-power turbine to propel the ship, and in most cases a non-regenerative system free from complication and added weight would be stipulated.

Hovercraft

Strictly speaking, the word hovercraft is a proprietary term applicable only to Air Cushion Vehicles built by or under licence from the Hovercraft Development

Ltd, and it should be reserved for that type of craft only. However, as in the case of the words 'Weirs' and 'Gwynne' used at sea for all feed and circulating pumps, irrespective of manufacture, hovercraft is now used indiscriminately to designate any type of vessel floating on a volume of air contained between its structure and the surface of the water, and for simplicity's sake the description will be confined to the marine type. The correct expression Air Cushion Vehicle, or simply ACV, is the generic term for many different makes of similar-type craft, but hovercraft seems so apt and descriptive that it will in all probability continue to be used instead of the more cumbersome technical name. Also, attention might be drawn to the use of the word 'vehicle' in this context, not 'vessel', as one would expect to see in a book devoted to marine engineering.

The fact is that most ACVs are as much at home on land as in the water, but unfortunately, as far as I have been able to ascertain, they do not come under any single authority in the legal sense. They can be considered as being simultaneously ship, motor coach and aeroplane, and their structure and navigation must conform to the appropriate sections of the Air Registration Board, the Road Traffic Act and the Board of Trade, with minor stipulations laid down by the Admiralty for good measure.

Air Cushion Vehicles have been with us a scant dozen years: they are a new breed of craft on our coasts and rivers, the last in the long queue of ship types since the earliest navigators. Yet the air-cushion principle is not new, it antedates the diesel engine, the steam turbine, even the compound steam reciprocating engine. Sir Frederick Bramwell, a past-president of the Institution of Mechanical Engineers in the 1870's, claimed that in the 1850's he had seen a ferry boat on the New York–New Jersey run having a system of air injected into a series of cushion areas fitted below the bottom in an attempt to reduce skin friction by a system of bubble lubrication between hull and water.

Clement Ader patented a plenum chamber-type air-cushion vehicle in 1904 using a multi-cushion design, while side-wall craft were patented in Sweden in 1909. A type of side-wall vessel, the ram-air machine, was developed in the USA before World War I. It consisted of a flat-bottomed speed boat with shallow side walls and a form of swim bow. When under way, air was swept under the bow and compressed between two side walls to form the cushion. Again, Robertson Porter used a skirted ACV as long ago as 1911 and several other inventors have tried out various ingenious devices to enable a vessel to remain poised above the water on a layer of air, although without any lasting success, possibly because the internal-combustion engine, which alone could propel such an airborne vessel, had not yet acquired sufficient reliability.

The principle of the Air Cushion Vehicle is beautifully simple. Place a large paper bag flat on the table with its open end hanging over the edge. Lay a massive book squarely over the bag and then blow lightly into it, when the heavy book will rise from the table and float upon the bag, thus demonstrating once again in dramatic fashion the superiority of hot air over muscular effort. It will be noted that the air pressure required to lift the book is very low indeed, of the order of inches of water, and this can be supplied by ordinary low-pressure fans. The difficulty in using the principle lies in containing the pressure in such a way that the air envelope itself shall not constitute a drag in the water. The paper bag or its equivalent must therefore be replaced by some other form of air cushion, and this is in effect the very crux of the ACV design.

One of the earliest uses made by the air cushion was for transporting heavy goods about the shop floor on an air pallet, a flat, hollow, box-shaped structure with a series of holes in the bottom. When low-pressure air was supplied to this pallet through a trailing hose it was lifted slightly off the ground and could be pushed about by hand, even when heavily loaded. This air pallet required a smooth surface to work on and thus provide a minimum rise height or clearance, and in the more sophisticated types the air holes in the bottom were controlled by spring-loaded valves opening inwards only when in contact with the ground.

In marine ACVs the air hose is replaced by one or more fans to continuously generate a cushion of air between the bottom of the structure and the surface of the water. This air cushion can be maintained by several different methods, but there is one primary requirement for all types of vehicle, they must be buoyant to begin with, i.e. they must be buoyant surface craft before they become hovercraft. Classification Societies define a marine Air Cushion Vehicle as a self-propelled vehicle which operates amphibiously or only over water and is capable of supporting at least 75% of its fully loaded weight, both while stationary and while under way, on one or more cushions of air which are continuously generated within the vehicle and which are dependent for their effectiveness on the proximity of the surface over which the vehicle operates.

The ACV must therefore be built as a floating, rigid structure capable of holding itself together and supporting its equipment, machinery, passengers, crew and cargo as a buoyant vessel without collapsing. The bottom of this structure is fitted with air trunking, either rigid or flexible, and/or flexible skirts to contain the volume of air supplied by the fans. Some ACVs, the side-wall craft, could be considered as intermediate between purely buoyant and free-flying craft, and they are possibly the oldest type in use. Not unlike a twin-hulled vessel decked over, these side-wall ships float on a pair of fore and aft

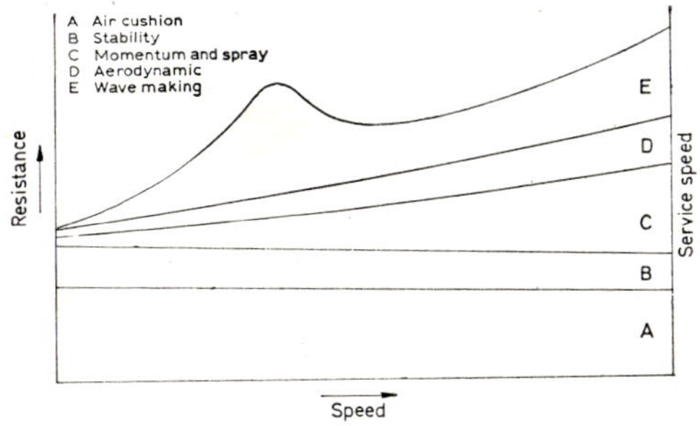

74 Power-consumption curves for ACV

trunks, either rigid or inflatable, the hull bottom between the two 'floats' sloping downwards towards the stern. At speed air is rammed below the bottom between the side walls and is compressed to give lift to the craft.

Other types use the plenum chamber, an open inverted box formed by the hull bottom and skirting on the periphery, the plenum chamber (i.e. chamber under pressure as opposed to a void space) being kept charged with air from the fans.

One of the best-known types was developed by Christopher Cockerell in 1958, using inflated skirting and an air curtain. Here the fans discharged directly into the skirts to keep these inflated and provide initial buoyancy. From these, through prepared holes, the air escaped in a series of jets round the whole periphery as a curtain of air to form a seal between ship and water, the air jets being directed downwards and inwards to acquire a certain energy to help the lift. The inflatable skirts and associated jets are made of light rubberised fabric.

Methods of propulsion and steering are as varied as the means of retaining the air cushions. In side-wall vessels, being semi-waterborne to start with, normal marine propellers and steering are usual, whether by rudder or swivelling propeller, and light diesel engines provide the motive power. In free-flying craft, aeroplane-style propellers driven by gas turbines, either direct or through layshafts, are more convenient. In this case the screws are mounted on pylons that can be either fixed or swivelled for steering and reversing. Alternatively, air rudders may be used.

In many cases the gas turbine or diesel engine drives both propeller and fan from one common line of shafting, the power ratio of propulsion to lift being in the region of 3:1.

Yet another means of propulsion and steering is by air jet alone, the air being allowed to escape through jets either at the stern or from the sides, for steering. Here it might be opportune to add that in the case of two or more fans being used to provide the lift, their intakes are fitted with individual non-return shutters to seal them in the event of fan failure.

Sea-keeping qualities in air-cushion vehicles are claimed to be at least as good as buoyant vessels of equivalent dimensions. Unlike the aeroplane, the cushion vehicle is inherently safe, safer in fact than the normal marine craft inasmuch as it disposes of two methods of flotation: air lift and displacement buoyancy, a gilt-edged belt-and-braces insurance against misadventure. This does not mean, of course, that an ACV is immune to any form of disaster, although capsizing is possibly the last mishap one would expect of a vessel shaped like a beamy pontoon. On the other hand, ACVs can travel at up to 60 knots and more, and in making a turn, with no lateral area presented to the water as in a normal vessel, there is always the tendency to skid with the danger of collapsing the air-retaining skirt and plunging sideways into the water: the nautical equivalent of a burst tyre on a road bend.

Economically, the air-cushion vehicle ranks with the hydrofoil in operational costs, just slightly below the cost of air travel but still very much higher than with the conventional passenger vessel, and to date it is confined to short-haul high-speed traffic such as cross-channel and lake-ferry service. Old shell-backs need have no fear that these high-speed vehicles will take over from the orthodox cargo and passenger ships within the foreseeable future. They will take their proper place in the Merchant Navies of the world but they will not replace them.

13 Steam boilers

The history of the steam engine is so intimately bound up with the history of the steam boiler that the one would be incomplete without the other. By definition the steam engine implies the use of a boiler and in fact one is the complement of the other, they form the two halves of a complete unit. Taken singly, neither engine nor boiler could show good reason for its existence, together they form the perfect partnership, with the engine as the dominant partner. The engine always commanded better design, more meticulous workmanship and greater care and attention as well as being housed in a well-lighted and airy engine room, while the boiler, as befitted a poor relation, was tucked away out of sight in a dark and smoky stokehold. After all, apart from the engineers, who wanted to see the boilers? In the heyday of the Clyde steamers a special gallery was built round the top of the engine room and it was almost a ritual for the passengers to lean upon the polished teak rail and gaze at the 'wheels going round'. Whether diagonal or oscillating, the engines were always worth looking at, and they always attracted the crowd—but nobody asked to see the boilers. In the mail-boats, a tour of the engine room was always a recognised event in the cooler weather, with the watchkeeping engineer in an agony of apprehension lest some of the passengers should get themselves wrapped round the shafting, but the visit always stopped short of the boiler room. Even in an oil-fired stokehold, clinically clean and bright as a new pin, there was nothing to attract the attention, nothing moving. It was the engines the passengers had come to see, especially steam reciprocating engines which have a magic all their own. Thus, to most people, a ship's boiler, if they thought about it at all, was simply a big tank that produced steam.

The steam boiler then, considered as a pressure containment vessel and not as just a steaming vat, came into the world together with the first steam pumping engine at the beginning of the eighteenth century. The boiler supplied with Newcomen's engine, as depicted in the illustrations, shows a rudimentary haystack tank boiler with a clamped manhole lid supported on masonry and heated from below. As this boiler supplied water vapour only to the engine and not steam under pressure, the boiler scantlings would not vary to any considerable degree from those of the dyer's or brewer's vat, i.e. it could be built and erected by local tradesmen. It was not a pressure vessel as we understand the term.

Fluid pressure was unknown to the ancients, hence the number of beautiful and elaborate aqueducts throughout the Mediterranean lands, spanning wide valleys or deep gorges, to carry water to the towns. This water supply could have been conveyed just as easily at a fraction of the cost by leaden pipes or hollowed-out tree-trunks under a static head, in keeping with the fashion existing for domestic supply, but for the idea that water could not possibly run uphill, even

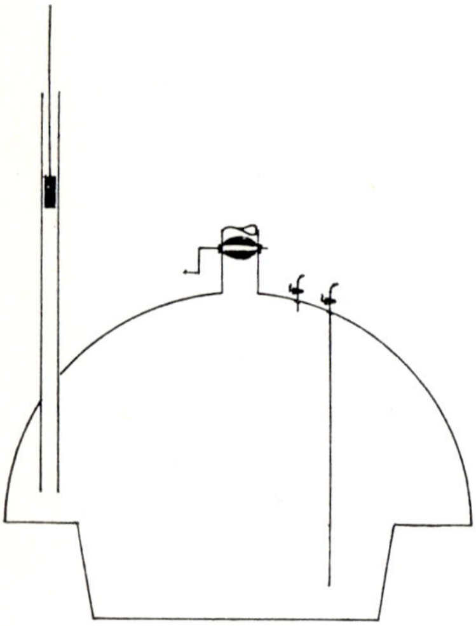

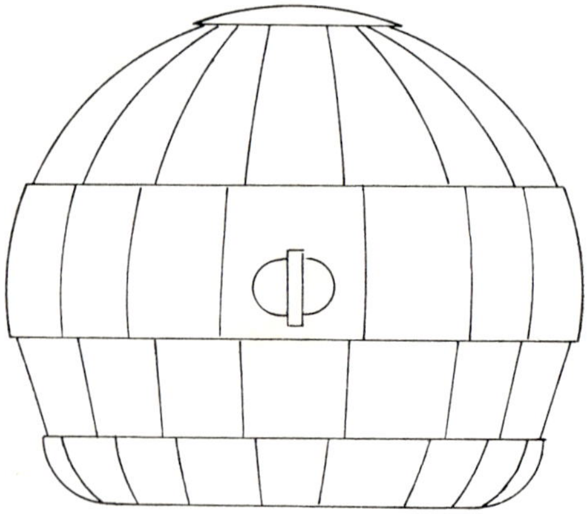

75 *Left:* Newcomen type boiler showing pressure-gauge tube and water-level cocks
76 *Right:* Early haystack boiler

inside a pipe. The concept of pressure in a pipe or in a boiler never entered engineers' heads until they were faced with a problem that demanded pressure, and it was James Watt who realised how it could be used.

Once the steam engine had established itself on pressure steam, even if only at 5 lb over the atmosphere, a vast number of boiler designs appeared on the market, some of which reappeared in high-pressure water-tube boilers a hundred years later, when more sophisticated materials and tools became available. Most of these designs fell by the wayside, they were not practical and took little account of the dangerous possibilities of steam under pressure. Some designs, however, were considered good enough to leave their mark on industry and attained a modest reputation for efficiency. The Jordan boiler, for instance, consisted of a number of 8-inch-bore cast pipes about 5 feet long set vertically in a sheet-iron casing. These pipes were connected at their lower ends to the feed pipe and at their upper ends to the main steam line, in effect a type of multiple-flash boiler, placed in the hearth of a wide furnace. Another type was Harrison's cast-iron boiler, made up of elements like a series of radiators, each element composed of a number of hollow cast-iron balls joined together. These elements were set at an angle of 10° to the horizontal, with the feed pipe at the lower end and the steam tapped off at the upper end.

Another cast-iron boiler, the James' ring boiler, consisted of eight to ten large hollow-cast rings, of 6 inches square section, placed together to form a cast-iron cylindrical furnace inside a casing. Again the feed and steam pipes were connected to each ring, respectively at bottom and top, and the fire bars were placed at half-height inside the series of rings.

These cast-iron boilers may have been suitable for land engines but most certainly not for ships' use with salt feed, and the earliest steamships built on this side of the Atlantic favoured the haystack form of boiler for at least the first two decades of their existence. This 'haystack' design, in fact, lasted for over 100 years as the paddle steamer *Mountaineer*, built in 1910, had boilers of this shape. A short description of an actual boiler therefore might not be out of place at this juncture.

The boiler in question was built for one of the early Thames steamers of a shape not unlike that of the Newcomen boiler, i.e. hemispherical top on a roughly cylindrical base. It was 14 feet in diameter at its greatest width and stood about 12 feet high. The shell consisted of four tiers of plating, each tier with twenty-one plates, and a top and bottom crown plate. The plate thickness varied from $\frac{3}{8}$ inch for the upper tiers and the crown plate, to $\frac{7}{16}$ inch in the waist and $\frac{5}{8}$ inch for the bottom plating resting on the foundation masonry. The plates in each tier were exactly alike and were riveted together by single rivets at 2-inch pitch. This boiler was completely self-supporting without stays of any sort and was without a doubt a triumph of the boiler makers' skill. The bottom of the boiler was dished inwards to form the firebox, and there was a manhole in the dome, with safety valve and $1\frac{1}{2}$-inch feed-cock opening.

By the time the steamboat had established itself and had ceased to be an experimental vessel, i.e. when ship-owners were suggesting to shipbuilders that ships be run for profit and not for research, it was appreciated that the haystack boiler took up too much room in the machinery space for the energy it supplied. The trend in boiler design switched to rectangular sections, big iron boxes that could be neatly stowed in the rectangular hold space set aside for the machinery. This design, however, brought considerable disadvantages in its wake as soon as pressures of over a few pounds were required. Cylindrical or dished plates in a pressure boiler are reasonably self-supporting, flat plates are not. Hence a boiler built as a parallelepiped had to be closely cross-stayed in all three planes. This form of construction raised additional problems of access. With the old haystack

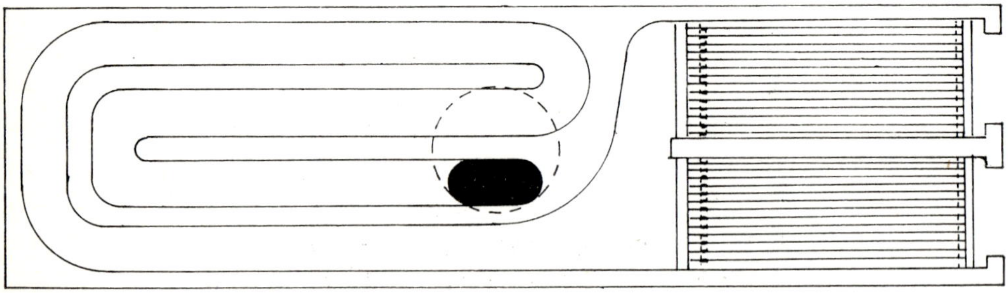

77 Rectangular boiler, horizontal section through furnaces and sheet flues leading to uptake

boiler, it was a simple matter to move around inside for scaling and maintenance; in the 'Square' boiler it was almost impossible: maintenance went by default. Thus boilers of this type soon either scaled up completely or their bottom plating corroded away, but they had one point in their favour: they had internal firing. The old 'haystack' boiler was essentially a plain tank, like its predecessor Newcomen's alembic boiler, supported on masonry and fired from below with most of the heat escaping up the funnel. The rectangular boiler had the furnaces placed within the water space, with their flues conveying the heat from the exhaust gases to the remainder of the water.

However much we may decry the design of a box-shaped pressure vessel, it was the forerunner of the tubular boiler, the present-day Scotch boiler, which was the keystone of tramp machinery for over fifty years.

The original low-pressure rectangular-shaped boiler had Gothic furnaces, i.e. of rectangular section with semi-elliptical tops, extending into labyrinthine sheet flues scarcely wide enough to allow access for sweeping. All flues met at a common breeches fitting at the base of the funnel. These flat-sheet flues were extremely vulnerable to collapse and the maximum permissible pressure in a boiler of this design could not exceed 5 lb. A typical box boiler would have the following characteristics: Length 25 feet, width 9 feet, height 12 feet. Three rectangular furnaces per boiler each 7 feet 6 inches long by 2 feet 6 inches wide leading into flues 18 inches wide, with 5-inch-wide water spaces between them. Length of flue pipes from furnaces to funnel about 75 feet. Weight of boiler with water about 50 tons. Life of boiler if made of wrought iron: three years.

In passing, it could be pointed out that in the case of these large low-pressure boilers there was as much pressure on the bottom plating when cold (due to weight of water) as on top of the boiler under steam. However, in spite of their serious maintenance troubles, i.e. constantly sweeping flues and removing incrusta-

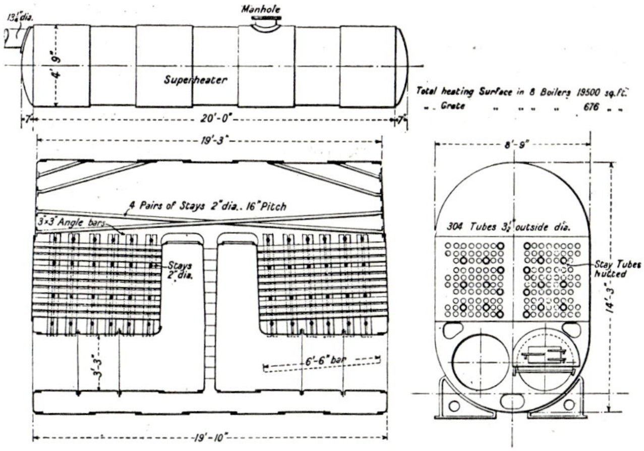

78 Egg boiler (1870)

tions from the water spaces, flue boilers continued to be built until well into the 1860's.

By mid-century, when engineers had evolved a technique of rolling wrought-iron tubes, the tubular boiler came into being, still in the shape of a parallelepiped, but now with the furnaces leading into a common combustion chamber and with return tubes leading the gases back to a smoke box on the front endplate, thence to the uptakes.

With the inevitable increase in working pressures, boiler shells, furnaces and their combustion chambers took on a rounded shape, though still retaining flat endplates. The first major step in that direction brought out the 'egg' boiler, oval in section with rounded top and bottom plates, but with flat side and end-plates, each circular furnace led into its own combustion chamber, with separate nests of tubes. This type of boiler usually carried an extension to its steaming space called the superheater. It consisted of a small-section circular drum, the length of the main boiler, and connected thereto by a wide throat at the top of the steaming space. This superheater was situated in the uptake passages below the funnel and served to dry the steam on its way to the engine.

By the 1880's the cylindrical Scotch boiler, single- or double-ended and more or less as we know it, had established itself for all types of ships. Most of the double-ended units still retained separate combustion chambers for opposite furnaces, but eventually the design crystallised into a type with, say, three combustion chambers for six furnaces. Apart from minor differences in furnace design and shape of combustion-chamber tops, the Scotch boiler of the 1880's was similar to that of the 1940's.

Until the 1880's plain cylindrical furnaces were invariably used, but in 1877 Fox brought out his patented corrugated furnace to take up expansion with the boiler under steam and towards the end of the century Scotch-boiler furnaces were rolled corrugated to Fox, Morrison or Deighton corrugation profile. By this time, too, the 'superheater' drum or steam dryer had disappeared to make way for the genuine combustion chamber, smoke box or fire-tube superheater, whereby the steam was heated above the saturation temperature, not just drained of entrained water.

For a proper understanding of the early Victorian marine engineer's difficulties with steam, we must examine the working conditions of the first tank boilers, either flue or tubular. With steam pressures ranging from a modest 3 lb in 1820 to 20 lb in 1850, it was not an easy matter to control either water-level or steam pressure with the primitive appliances then in use. Originally, a pair of test cocks were used to gauge the water-level, the cocks being placed about 12 inches apart, and the lower cock fitted with a tube extending well below the water surface. This idea works well with low pressures and with the water surface at rest. At sea, the movement of the ship causes violent surging of the water and the cocks are unreliable. In later models a guided float inside the boiler, attached to a piano wire through a bush in the boiler top, operated a pointer on the bulkhead and registered the water-level. This also was unreliable: the float could stick or the wire could break or jump off its guide pulleys.

Again, there were no adequate means of measuring the pressure in those far-off days. The first dial pressure gauge, invented by Sydney Smith in 1847, was not in general use until the late 1850's, and before that date the mercury column had to make do. This was a short U tube screwed directly into the boiler shell and partly filled with mercury. A small steel float carrying a pointer against a scale registered the steam pressure as the mercury level rose or fell.

There was no trouble in feeding boilers working in the lower ranges of steam pressures. The air pump worked by the main engine discharged most of the jet-condenser-spray water overboard, but part of this was bypassed into a tank on deck which served the dual purpose of boiler-feed tank and wash-deck service. From this tank feed flowed by gravity into the boiler through a control cock.

On the other hand, it was quite a problem to empty one of these large low-pressure boilers. With steam pressures below 5 lb. they could not be blown down into the sea, the pressure head outside was greater than that within the boiler; they had to be drained off into the bilges and thence overboard by hand pump (assuming the main engine to be stopped). Nor was it always convenient or even safe to discharge some 40–50 tons of scalding water into a ship's engine room. In practice it was necessary to freshen up the boiler contents to counteract the increase in salinity caused by evaporation. With pressures of 10 lb and over, this was done either by a constant blow-down, topping up with extra sea-water feed to maintain the water-level: or else through a routine blow-down of all boilers twice a watch and fill up with cold sea water. In either case it involved a very considerable waste of heat overboard.

With lower pressures, where the boiler could not be blown down against the head of water outside the ship, and where obviously a constant blow-down could not be discharged into the bilges, a 'brine pump' or more correctly, a 'brining pump' was used to draw water from the boilers and discharge it overboard. This pump, a form of negative-feed pump, was driven off the main engine shaft by the same means as the bilge and air pumps.

Two other fittings on those old tank boilers must be mentioned: the deadweight safety valve and the atmospheric valve. In the earlier models, the safety valve had no escape pipe, the valve simply blew out into the stokehold to the discomfort of all personnel in the vicinity. There was no control over the number or type of weights fitted and anything could be hung on to the valve spindle to increase the pressure. As for the atmospheric valve, this was a most important item: a boiler built to withstand a maximum of 5 lb steam pressure would not support an external pressure of nearly 15 lb, it would just collapse. Hence a valve opening to the atmosphere to break the vacuum as the boiler cooled off.

The marine engineer's biggest problem throughout the whole of the nineteenth century was to prevent priming in his boilers. Priming in a boiler under steam is caused by severe agitation of the water surface, producing foaming in the steam space and carry-over into the engine cylinders. Severe cases can result in cracked pistons or cylinder covers and will upset the metallic packing in the HP piston and valve glands. It will not be found in a well-maintained boiler working on good clean fresh-water feed, priming is basically a phenomenon of sea-water feed and heavy internal lubrication, i.e. of the jet condenser and the tallow kettle.

When a boiler has been steaming hard for several days on salt feed, the production of steam will have caused the density to rise to a dangerous level, to the point where salt may deposit on the heating surfaces and bring about furnace collapse and explosion. To prevent this, the boilers must be partly blown down regularly and topped up with more salt feed, and after a few days be completely emptied and filled up with a fresh supply of ordinary sea water. Hence, under steam, the boiler nearly always has a dangerously high density.

Now in the days before mineral, or even vegetable, oils became available, tallow was the universal lubricant for any engine, both inside and out, and internal lubrication was necessary for the heavy, slow-running pistons and valves. All this grease eventually found its way into the jet condenser and was pumped back into the boiler, to form a heavy scum on top of the water surface. This greasy, frothy mass lying on water of a high salinity formed ideal conditions for priming—the whole mass would 'boil over' and fill the steaming space with froth and water in suspension. An idea of the effect of priming will be obtained by comparing a saucepanful of boiling water and one of boiling milk. The one will only bubble the other will boil over.

A further cause for priming, if that were needed, was the prevailing habit among engineers to put potatoes and oatmeal into their boilers to prevent leakage in the seams, and to coat the internal surfaces of the plating with either tallow or graphite.

The Scotch boiler was the ideal other half of the reciprocating engine unit as its maximum design pressure coincided with that of the cargo-ship triple-expansion engine, usually at 180 lb but sometimes up to 220 lb, occasionally as high as 250 lb. But it was, to us, still a medium-pressure boiler, not suitable for the higher pressures required for turbine machinery. It was also a very slow boiler to get under way: twelve to twenty-four hours, and although this could be reduced to six hours in an emergency (getting up steam on receiving a hurricane warning, say) the boilers would never be the same again.

These points were realised early on in the nineteenth century, and many attempts were made to put a water-tube boiler on to the market, but without success. The nineteenth-century marine boiler was conditioned to run on salt-water feed and it had to be of the tank type to allow of internal descaling; a water-tube boiler just could not run under these conditions. In spite of this, however, the water-tube principle was introduced into tank boilers at about the same time as smoke-return tubes became fashionable, somewhere about mid-century. The first boilers of this type were the haystack boilers not to be confused with the simple 'haystack' shell boilers of the early steamships. This later type of haystack boiler, which attained a modest popularity in the latter half of the nineteenth century, and even into the twentieth, among the river paddle steamers, was a large, vertical, cylindrical boiler with a domed top and it had a dry bottom in way of the ashpits, with a wet uptake.

It combined the large capacity and internal accessibility of the Scotch boiler with the rapid steaming characteristics of the water-tube boiler. Its main disadvantage lay in judging the water-level. With a free-water surface of 10 feet in

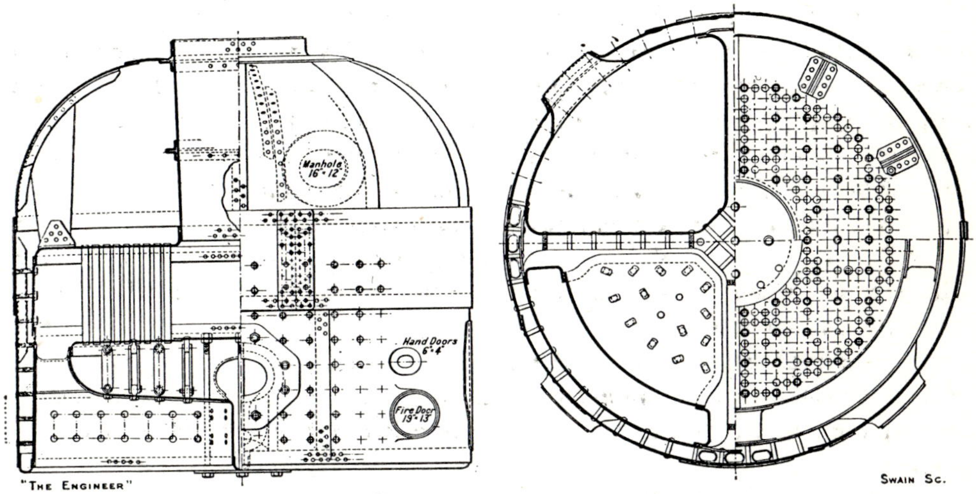

79 Revised form of haystack boiler with water tubes (1910)

diameter surging about from side to side with the motion of the ship, and with the rapid evaporation from the water tubes, the water-level was determined by experience rather than from direct observation, a form of dead reckoning in fact.

The boiler had an outer and an inner shell, the top of the latter forming the upper tube plate. The bottom tube plate with the furnace crowns formed a large saucer-shaped pan with water pockets separating the four furnaces set round the shell at 90° to each other. These water pockets retained the sediment deposited from the circulating water, and which could be removed by hand-holes in the shell. The water tubes, about 3 inches bore, were fitted between the top of the furnace pan and the top of the inner shell. In spite of its rather clumsy appearance, this haystack boiler was occasionally worked at 120 lb pressure.

The earliest true water-tube boilers appeared in the 1860's and already they showed the characteristics common to all their successors: a steam drum at top and a water drum at bottom connected by a series of water tubes in direct contact with the products of combustion, with a minimum water space. The Benson and the Belleville boilers were similar in principle, simple horizontal steam and water headers connected by a number of vertical tube elements, each element consisting of straight lengths of tube connected by return bends to form a series of hairpins. In the Benson boiler there were return bends at each end of the tubes, which could be removed for scaling. In the Belleville boiler the tubes were shaped

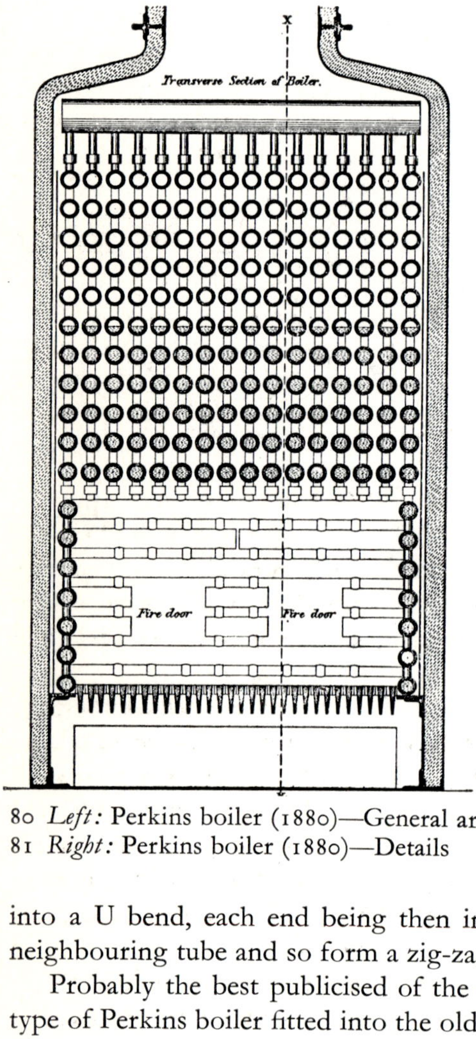

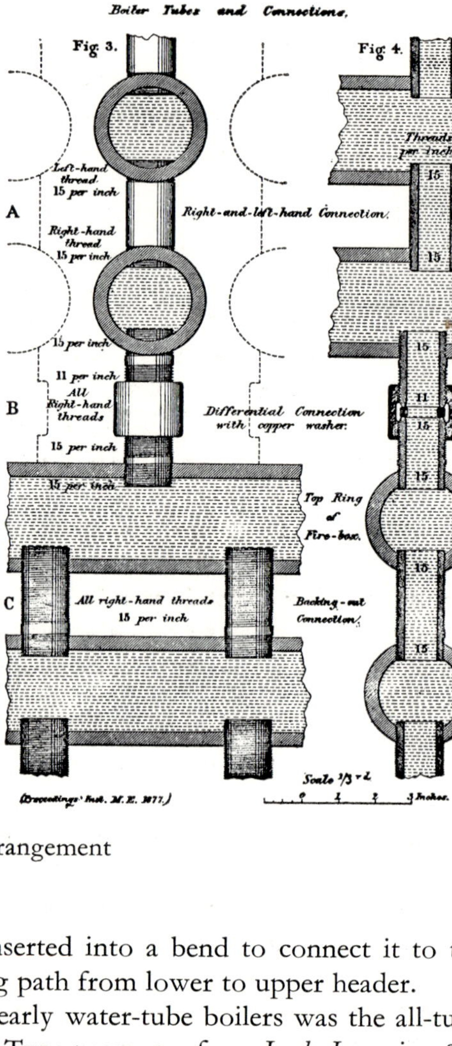

80 *Left:* Perkins boiler (1880)—General arrangement
81 *Right:* Perkins boiler (1880)—Details

into a U bend, each end being then inserted into a bend to connect it to the neighbouring tube and so form a zig-zag path from lower to upper header.

Probably the best publicised of the early water-tube boilers was the all-tube type of Perkins boiler fitted into the old Tyne passenger ferry *Lady James* in 1880 and subsequently renamed *Loftus Perkins*. This had the safety valves adjusted to 450 lb, quite a high pressure even today, but an astonishingly high pressure for a boiler in 1880, when the prevailing 'high pressure' meant 60 lb.

This boiler was surprisingly modern in conception, and consisted of a main bank of tubes above the hearth with water walls round the furnace, the steam-collector drum being very small in diameter. The tubes, with blanked ends, were connected to each other by screwed thimbles with right- and left-handed threads, and to the water walls by screwed thimbles jointed together with copper washers by a running coupling. Tubes were 2¼ inches nominal bore, their thimbles being 1 inch in diameter or thereabouts.

This type of boiler had one serious defect, it was impossible to descale the tubes without dismantling them and rebuilding the boiler. Even using good clean town water as feed, the tubes would soon scale up at the high rate of firing required.

The water-tube boiler only came into its own towards the turn of the century when Babcock and Wilcox brought out their straight-tube header boiler in 1895; straight tubes because they still had to be swept by hand. This boiler was formed of a large steam drum connected to a small mud-drum at the bottom by a series of sinuous steel headers, with a corresponding series of headers at the back of the furnace, and with the straight water tubes connecting the two sets of headers. These tubes sloped upwards towards the back at an angle of 20°, and from the top of the back headers a series of horizontal tubes led into the steam drum. The tubes were of the order of 2¼ inches diameter, with two rows of 4-inch tubes at the bottom in way of the hottest part of the furnace, the larger diameter opposing the tendency of the tubes to bow away from the heat.

This type of boiler was extremely rugged and soon became popular for all steaming requirements, from large passenger-liner main boilers to motor-ship donkey boilers or for any purpose requiring moderately high pressures.

A completely different type of boiler appeared shortly after the turn of the century: the Yarrow three-drum small-tube type. This was in the shape of a triangle with the steam drum at the top and a water drum at each lower corner, the triangle forming the combustion space, the steam and water drums being connected together by a number of rows of small-bore tubes. The superheater was inserted in a bay inside the tube nest on one side of the boiler, in the uptake path, the tubes being connected to inlet and outlet headers. The air heaters and/or the feed economisers were installed beyond the evaporating tubes in the uptakes. This design eventually led to the five-drum type, with one steam drum and three water drums, connected as before in arrow-head shape ᐱ, with a divided superheater drum between two water drums, the hairpin return tubes of which were sandwiched between two banks of evaporating tubes.

In the earlier water-tube boilers the superheat temperature was determined by the steam loading, the design did not permit of any alteration other than by mixing with saturated steam or by an attemperator, i.e. running the superheater outlet pipe through the steam drum. The latest models now have independent means of controlling the superheated steam temperature, either by special bypass louvres in the uptakes, or by placing the superheater elements in a separately fired furnace, adjacent to the main one. Another refinement is in lining the furnace walls with water tubes, the water-wall type, thus absorbing as much of the radiant heat as possible, even to the extent of welding all the wall tubes into a common steel water wall by means of longitudinal fins on each side of the tube—the so-called membrane wall.

These sophisticated boilers are obviously intended only for the most advanced machinery conditions: high-pressure, high-temperature steam, and they demand a degree of upkeep and maintenance found only in the largest passenger steamers or in the super tankers.

The Babcock and Wilcox and the Yarrow boilers dominated the water-tube boiler field in the Merchant Navy during the first few decades of the twentieth century, but they have since been joined by a number of rivals, among which may be mentioned the Foster Wheeler twin-furnace boiler, the Johnson two-drum boiler and the La Mont forced-circulating boiler. In general, the trend in high-pressure boiler design is for a two-drum arrangement with tubes in a D contour forming the furnace water walls and enclosing the superheaters between nests of tubes.

Several exotic all-tube boilers have appeared in recent years giving high pressures and high temperatures for a very light-weight structure, using mechanical means to circulate the water at speed through small-bore tubing. In the Loeffler boiler saturated steam is generated in the water drum by superheated steam bled off from the main outlet to the turbine. This saturated steam is then circulated by pump through a radiant superheater in the furnace itself. The Benson boiler, already mentioned in the 1860's, reappeared as a straight-through flash boiler with forced circulation, and was one of the lightest construction boilers ever marketed. The La Mont boiler, already mentioned, uses a steam turbine-driven pump in open circuit with the boiler which forces the water through the furnace tubes in a strictly defined distribution by means of resistance nozzles at the entrance to each circuit.

Another type of boiler, a low-pressure auxiliary boiler, works rather on the principle of an evaporator, or like the Loeffler boiler, in that it uses the heat from high-pressure superheated steam to generate low-pressure steam from feed water. This boiler, called a steam-steam generator, works in series with the main high-pressure boilers to provide steam for hotel services and oil heating, etc., at about 100 lb.

Steam boilers are not by any means confined to steamships, most motor ships have donkey boilers to supply steam for a great variety of purposes. Many motor ships have steam-driven auxiliaries and all passenger ships require steam for heating, cooking and laundry services. Motor tankers need steam for tank cleaning and heating, pumping, etc., and in fact in any motor ship there are many little ways in which steam can be used. Again in all motor ships a very large amount of heat is discharged up the funnel, heat that can be used to raise steam and provide useful work. But, and this is a most expensive but, as many ship-owners realise, to raise steam requires a boiler with all its ancillary equipment: feed pumps, condenser, circulating pump, oil-fuel burning equipment (for harbour use), special fire-fighting equipment, *et al*, but above all, the boiler needs an attendant to look after it in commission and somebody to scale and overhaul it when empty, and this sort of labour costs money. On an average all-electric cargo motor ship, the steam gained from the waste heat can be put to good use in the galley or in heating the accommodation, but at an extremely high price in extra equipment: heating and cooking by electricity would involve little additional cost in power, with no special apparatus required.

During the whole of the nineteenth century, and well into the twentieth, the marine engine depended exclusively upon coal for its very existence, and it was not until 1894, when the *Baku Standard* made the first crossing of the Atlantic burning oil fuel in her furnaces, that the marine engineer freed himself from the domination of the stoker. This ship, a new tanker of 3705 tons, had the oil sprayed into the boiler furnaces by means of high-pressure steam jets and it was calculated that it took over one ton of steam for every ton of fuel so burned. To make up all this fresh water in the evaporators, an extra $1\frac{1}{2}$ tons of fuel daily had to be burned, adding considerably to the already high cost of oil fuel. To further complicate matters, the early oil-burning steamers faced the same problems as the early motor ships: lack of bunkering facilities. To offset these drawbacks, however, the ship-owner saved considerably in stokers' wages and profited by a huge increase in cargo space through omitting the coal bunkers, while passengers were spared the horrors of coaling in a steamy tropical port.

In the early oil-burning ships the boilers were usually arranged to burn either coal or oil. In the latter case the fire-bars were left in position and covered with a sprinkling of broken fire-brick and clinker. The oil burner at the furnace mouth was a fitting comprising a hand-operated needle valve placed at the front of a pair of cast-iron nozzles in series. The oil, controlled by the needle valve, was injected into the throat of the steam nozzle and blown into the furnace, entraining a supply of air through the secondary diffusing nozzle. The fuel oil, at 60 lb, and the steam at boiler pressure, were circulated through heating coils in the smoke boxes, while the combustion air was drawn in from the stokehold through a long U tube in the furnace ashpit.

Present-day steam boilers are built under such exacting conditions of design, materials, construction and testing that a boiler explosion in the Merchant Navy is now happily a thing of the past, even if we extend the term, as do the Board of Trade, to include explosion in an evaporator, air receiver, rupture of a steam pipe or failure of a stop valve. The sudden and violent release of high-pressure steam, air or gas into a crowded engine or boiler room invariably causes damage to personnel and machinery and places the ship and her crew at risk.

To prevent such an unfortunate happening, every item that goes to make up a merchant ship's machinery and boilers is built to a recognised high standard under the supervision of a Classification Society or the marine department of the Board of Trade, sometimes both, who carefully scrutinise and check the drawings, manufacture, assembly and testing of all components, individually and in relation to each other. These authorities act independently of ship-owner and shipbuilder, and the finished product is therefore proof, as far as human ingenuity and knowledge can make it, against disaster under service conditions. It will most certainly be guaranteed against any skimping in materials and corner-cutting dictated by greed or false economy.

But conditions were not always as straightforward and clear-cut in the old days. We have already described the terrible accident in 1847 to the s/s *Cricket*, of the Halfpenny Steamboat Company operating on the Thames, when the starboard boiler back endplate blew out and the boiler took off through the cabin, killing seventeen passengers and seriously wounding sixty others.

One of the worst boiler explosions in the Royal Navy happened in 1876 on board HMS *Thunderer*, a twin-turret ship built in 1872 and the last capital ship to have square boilers. This ship used steam at 39 lb supplied by eight boilers in two stokeholds, and the accident happened to the forward starboard boiler in the after stokehold through failure of the front endplate, due to excessive pressure. The explosion killed fifteen men instantly and seriously wounded seventy others, thirty of whom subsequently died of their injuries, and the enquiry revealed that the safety valves on the boiler in question had seized in their seats through corrosion.

In another vessel, the *Elbe*, the main steam pipe burst in 1887 killing nine men, while in the German belted cruiser *Brandenburg* in 1894 the main steam pipe on the starboard side burst, killing forty-one men and seriously wounding many more.

Possibly the most horrible of boiler accidents happened in the US gunboat *Bennington* in 1905 when a furnace in one of the Scotch boilers collapsed and blew out, the sudden reaction jumping the boiler off its cradle and fracturing the steam pipes attached to it. Over sixty people were killed in this holocaust.

82 Examination of a boiler after mishap (1876), with colza lamp as only available light

Many of these explosions happened more than a century ago, and it is now almost impossible to discover exactly what caused some of these steam pipes to burst, if in fact 'burst' is the correct verb. In those days pipes were made of copper with riveted or braised seams and flanges and with their relatively small diameters and resistance to corrosion there should have been ample margin of strength. Two reasons suggest themselves: water hammer and expansion joints, both of them features of design. Water hammer, if severe, can shatter a cast-iron stop valve into small pieces or split a steam-pipe bend wide open, but its action is due to a shock wave rather than static pressure. In the old days of multi-boiler installations with cast-iron valves and Tee pieces and with a few hundred feet of stone-cold steam piping connecting the boilers to the main engines, it was a delicate art to level up all boilers under steam without shaking a load of soot off the boiler tops through water hammer in the piping.

The old-type expansion joint was an extremely dangerous fitting to have in any ship. It consisted simply of a length of straight steam piping let into a junction piece through a stuffing box, there being nothing other than the stiffness of the main piping to prevent the pipe end blowing out. After a number of accidents had drawn attention to this defect, a braised collar was fitted near the end of the pipe, and the gland studs were extended to pass through bolt holes in this collar and thus with their locknuts prevented the pipe from blowing out.

Probably the greatest contribution towards safety in engine and boiler plant was the introduction of electric lighting on board ship. Heretofore engineers, like coalminers, had had to work in a perpetual gloom in the engine room, relieved only by such daylight as managed to filter down from the skylights far above, and by the odd hurricane lamp or slushlamp hung about the engine at strategic points. Once the machinery had been installed and the engine room decked over, neither engines nor boilers could ever again be examined dispassionately as a complete unit. With the existing lighting available, individual components only could be inspected, and then only after a fashion.

In the boiler room it was worse, there being no glazed skylight or paddle hatch, only a narrow fiddley grating round the funnel. Here, inspection of the boiler tops, if inspection was ever made, could not be more than a cursory affair by the light of a tallow candle that would certainly not reveal corroded studs or wasted rivets. Inside the boiler, amid a maze of cross-stays and tube nests, any examination made by a one-candle-power light must inevitably be suspect: it would have to be a gross defect indeed to be noticed in such circumstances. Any minor flaw would pass unobserved, to develop into a serious hazard before the next boiler inspection became due.

Electric light, both fixed in the boiler room and by trailing leads suspended all round the inside of the boiler, shows up every part of the shell and fixtures in stark detail and allows a quick and correct assessment of the condition to be made and, more important, enables a careful repair to be carried out.

14 Auxiliary machinery

A favourite pastime for the bored passenger on a large luxury liner, after having savoured the exclusive if somewhat limited atmosphere of the navigating bridge, is to join in a conducted tour of the engine room to see the 'wheels going round'. There, with a little intelligent questioning and a lot of luck, amid the toil, steam and sweat, he may just begin to understand what exactly it is that pushes this monster ship through the water at over 20 knots. It will never occur to him, however, that most of the machinery he sees and hears around him will have no bearing whatever on propulsion, it is there simply for his comfort and convenience.

Modern passenger vessels, for a variety of reasons, have their machinery spaces subdivided into a number of compartments, each housing a specialised type of equipment, and in a really large steamer this suite of engine rooms can extend for three-quarters of the ship's length, usually spaced in a manner consistent with the geometry of the hull and the requirements of passenger accommodation, catering, etc. Thus, travelling from forward towards the stern we would pass first through the pump room and oiling station, refrigerating and air-conditioning engine room, the electric generating room and switchboard, the boiler room, main engine room, shaft tunnels and on a 'tween deck right at the after end, the steering flat. A stabiliser-engine room, or possibly two, would be situated somewhere in this machinery complex, and to further complicate matters there would be spaces devoted to such unromantic items as fuel- and lubricating-oil tanks, fresh- and salt-water heaters and sewage-disposal systems, each with its separate pumps.

All of this complicated and diverse machinery calls for considerable knowledge of mechanical, electrical, refrigerating and heating engineering with the requisite skills needed to repair and maintain it, and all of it is undertaken by the marine engineer in his stride as a matter of pure routine. But marine engine rooms were not always so complex. The early steamboats were built as sailing ships equipped with a paddle engine and boiler(s) to propel them, and all the other services that we moderns take for granted: heating, light, power and all the amenities of a sophisticated passenger ship just didn't exist. Even such elementary safety services as bilge, fire and wash-deck pumping were dealt with in sailing ships fashion by rigging the hand pump.

These conditions obtained during the first half of last century, and in cargo ships well into the second half; the *Great Eastern*, the biggest ship of the nineteenth century, was built with hand steering gear, and even as recently as 1895 a third-class cruiser was sent out to the China Station with manual-steering gear and hand capstan.

The earliest marine engines, being jet condensing, required an air pump to maintain the vacuum, and this was driven from the main shaft either by its own crank or indirectly by the piston rod or side lever. For several decades this was the only power pump on the ship, it was in fact *the* pump. Originally fitted to drain the condenser of injection water and condensate, plus a small amount of entrained air, and discharge it overboard, it eventually became a general-purpose pump for bilge, feed and wash deck service. By gagging down the overboard discharge valve the water from the condenser could be diverted to a tank on deck whence it could flow by gravity back into the boilers as feed. This tank also contained a handy supply of luke-warm water for baths, wash deck, fire and sanitary services. By shutting the sea injection cock to the condenser and opening the bilge connection (the original bilge injection) the bilges could be pumped dry, although after a while the large amount of ashes and coal dust inseparable with bilge water tended to form a dangerous sediment in the boilers.

In the course of time, and with the increase in boiler pressures, the incidental functions of the air pump were taken over by the feed pumps and the bilge pump, the feed pump drawing from the condenser hot well. Both these pumps were driven on a short stroke by the main engine, and being displacement pumps, they were single-acting and therefore of small capacity, but they required no difficult pump liners or awkward bucket rings, just a cored cast-iron body and a machined plunger. The bilge system was still a simple arrangement of single centre-line suction in the engine room. To drain the cargo holds the bilge sluices in the bulkheads were opened by the carpenter from the deck and the water was allowed to drain into the engine-room bilge to be pumped out by the single unobstructed suction pipe. In passing, this system of bilge sluices, to save the cost of bilge piping throughout the length of the ship, was still to be seen in some very old tramp ships as recently as the 1940's.

Another pump worked off the main engine in the old days was the brine or brining pump, which performed the services of the feed pump in reverse. As the boilers had to be blown down and periodically freshened by raw sea water, and as it was impossible to blow down with 5 lb steam pressure against a 10-foot head of water outside the ship, this water had to be pumped out and discharged overboard. The brining pump was unnecessary when boilers steamed at 10 lb and over and were never fitted on surface-condensing engines.

The most important of the engine driven pumps was undoubtedly the air pump as the main engine could not run without it; it was as necessary as the boiler itself. By contrast, the feed and bilge pumps made no direct contribution to the propulsion effort, but they played a significant part in the safety of the ship and her machinery, and at a fairly early stage in the development of the marine engine, the bilge and feed pumping were duplicated for manual as well as power operation. As early as 1845 the Admiralty required a hand-operated feed pump as standard equipment on board all naval vessels, but it was not until the 1860's that the ubiquitous donkey pump made its appearance in the Merchant Navy's engine rooms. This pump, the banjo pump, was the stand-by pump for all duties: boiler feed, bilge, wash-deck, fire, sanitary and condenser circulating when the main engine was shut down. When the main engine was being overhauled, the banjo pump also did duty as turning engine, with a rope traveller connecting its fly-wheel shaft with the turning-gear wormshaft.

This donkey engine consisted of a single steam-engine cylinder set on top of a cast-iron column and a water-pump chamber complete with bucket and valves at the bottom. The steam-piston rod and the pump-bucket rod were joined together by an open frame encircling the crankshaft at mid-height on the column, and the crank was driven by a connecting rod acting from the top of this frame. The function of the crankshaft was to provide an eccentric sheave to work the steam slide valve, and it carried a flywheel to help it over the centres. It would have been possible, of course, to work the slide valve from the piston-rod crosshead, to close when the piston came to the end of its stroke and work the pump non-expanding, but this would have entailed heavy steam consumption. The modern alternative in a simple pump: the shuttle valve controlled by an auxiliary valve actuated by the crosshead, was still a long way in the future.

The one great disadvantage of the banjo pump lay in its inability to equate running speed with pumping speed. Being essentially a single-cylinder rotating steam engine, it had a minimum speed below which it would stall, but this speed was still in excess of good pumping practice, especially when pumping warm water or bilges. The pump cylinder required time to fill, otherwise there would be lost motion in the bucket and irregular and noisy operation, but, even so, the clanking of the donkey pump in action has been aptly compared with the sound of a goods train falling down an embankment.

This type of pump continued in use as the only independent power pump for most cargo ships until the last quarter of the nineteenth century, when it was supplemented by the duplex pump, mainly for feeding. In this pump, really a pair of steam simplex pumps joined together Siamese-twin fashion, the piston rod of the one controlled the steam-valve gear of the other pump, the two strokes being phased 90° to each other in the SHM manner, with the valve cut off at roughly 50% stroke.

As steamers became larger and more sophisticated their bilge pumping equirements could no longer be dealt with by occasional use of the donkey pump, and some passenger ships used a form of hand pump fitted to the deck worked by a traveller chain from one of the winches. These vessels had, in fact, quite a choice of bilge-pumping facilities when at sea: by main engine-driven bilge pump, by donkey pump, by winch pump or, at a pinch, by direct suction from the jet condenser. But even as late as the 1880's, well-equipped cargo ships would be built with one single donkey pump in the engine room for all purposes, and even the *City of Rome*, one of the largest passenger vessels on the Atlantic run in 1880, with 8000 IHP in her engine room, could count on only three independent pumps; one each for feed, bilge and wash-deck services.

During most of the period covered by the paddle engine the exhaust steam was condensed in the jet condenser, where the bucket-type air pump combined the duties of circulating, air and hot-well pumps, and in its earlier days also as feed, bilge and wash-deck pumps. The surface condenser came into favour only with the arrival of the inverted steam-hammer-type screw engine, i.e. the prototype of our modern steam reciprocator, although it was introduced into the marine field as long ago as the 1820's, vide the *Victory*'s engines. It was revived about 1845 in some small coasters sailing out of Bristol and the Welsh ports using non-condensing engines and steam at 30 lb pressure. These boilers used fresh water to prevent scaling in their boilers and to avoid wastage they used a form of condenser, not to obtain vacuum but to save water. Originally exhausting the steam into a fresh-water tank, the idea was extended to include a special sea-water-cooled tank, then to a tubular condenser circulated by sea water. It is interesting to note that the circulating water was provided by a scoop in the ship's bottom.

The surface condenser made no headway until high-pressure boilers forced the marine engineer to use fresh-water feed. It was not that low-pressure boilers wouldn't steam much better with fresh than with salt water, the trouble lay with the large quantities of tallow introduced into the steam cylinders and valve chests to lubricate them. This tallow inevitably found its way, via the air pump, back into the boiler to form a spongy scum on the surface of the water inside the boiler, which in due course caused foaming or priming. An engine using a surface condenser would have to stop every so often to degrease the tubes, and that, plus the additional equipment required for fresh-water feeding, militated against the use of the tubed condenser.

However, by the 1870's this type of condenser was here to stay, and naturally it required a circulating pump to itself, a pump, moreover, that was of little use for any other auxiliary service. The surface condenser needed far more sea-water cooling than the old jet condenser, roughly in the ratio of 3 to 1, and hence the circulating pump was a most important component of the main engine, ranking equally with the air pump, whose function was now reduced to removing the condensate and any small traces of air brought in with the exhaust.

This circulating pump was needed only when steam was on the main engine, either in motion, warming up or shutting down, but it was not wanted in port, and for that reason it was generally driven by the main engine, together with the air pump, the feed and bilge rams and sometimes a sanitary pump, through beams worked off the crosshead. It remained a main-engine-driven pump for as long as steam tramp ships sailed the high seas, i.e. right into the 1940's, and it was a simple, straightforward and reliable bucket pump that could, in emergency, be used to pump bilges. In the larger steamers the condenser circulating was dealt with by a rotary steam pump, first introduced on land by Gwynne in 1849 but not in general use at sea much before the end of the century. This pumping unit consisted of a multi-vaned centrifugal impeller driven by a vertical steam engine, and in the larger pumps this prime mover was a two-cylinder compound engine. In extreme cases, the *Olympic* and *Titanic*, for example, the four main circulating-pump engines were as big as a tug's main engines.

A pump that has not been mentioned so far is the ballast pump, and oddly enough the ballast pump antedates the conventional ballast tanks. In the 1840's some steamers had their coal bunkers subdivided into small compartments, and as the fuel was progressively used up, each tank in turn was flooded from the sea through a system of pipes and cocks. Before bunkering, these tanks were pumped out through a 'special engine devoted to that purpose'. The true ballast pump, however, developed from the general-service or donkey pump and only came into its own when the cellular double-bottom tank was introduced to carry liquid instead of solid ballast.

Fire precautions in the modern vessel are served partly by one or more fire pumps in the engine room or adjacent spaces, delivering a continuous and copious supply of sea water through a network of piping to all parts of the structure; i.e. the pump and the piping arrangements are in the main stationary. In the nineteenth century, on the other hand, fire fighting was by a fire engine that could be carried or wheeled to the vicinity of the fire and there connected to a system of leather hoses dropped over the ship's side. This fire engine was equipped with a small flash boiler that could get up steam within minutes and start pumping upon arrival at the scene of the fire.

Feed heating, developed by Weir, was introduced in the last quarter of the nineteenth century, although tentative efforts had been made since the 'sixties to use the residual heat from the boiler gases for this purpose by fitting cast-iron economisers round the base of the funnel. At about the same time various other innovations could be noted: steam-steering gear, windlass and winches, mechanical telegraph between bridge and engine room, superheated steam, forced draught in the boiler furnaces and mechanical ventilation in the public rooms and passengers' quarters.

Passengers' comfort was not neglected throughout the years of steamship development, but somehow it failed to achieve any priority. As early as 1853 the steamer *Arabia* had underfloor steam heating and in 1876 an emigrant ship, the *Alfonso XII*, had a pair of steam-driven fans for ventilating the third-class quarters, but that was about as far as the ship-owner was prepared to go. Even in the large passenger ships there was little ventilation, and no running water, not even sea water, which was free. Hot and cold fresh water were supplied from the galley, and cold sea-water showers were available from the wash-deck hose only; but then the Victorian traveller was a hardy soul, who did not expect much and was seldom disappointed. He made no fetish of fresh air or cleanliness, and was not unduly fussy about quality in his food. In this he was fortunate as even the best of meat seldom lasted more than a few days in the ice box, while no galley was ever free of cockroaches.

Artificial lighting on all steamers throughout most of the nineteenth century took the form of oil lamps: on deck, in the engine room and throughout the accommodation. This was a safe and cheap method of illumination requiring no special apparatus and it endured in the modest cargo steamers at least until the 1930's. Until recently it was common practice in some ships to shut down the dynamo at sunrise and start up again at sunset, the watchkeeper in the engine room using a hurricane lamp on his rounds. However, as long ago as the 1860's the *Great Eastern* had been fitted out with a gas-generator plant to supply coal gas to illuminate the public rooms, although it is not clear if this plant was at all successful. And in 1872 the White Star liner *Adriatic* was fitted out to burn gas in the saloon and the engine room. This installation, which fed 300 gas burners, consisted of three retorts feeding into a storage gasometer, one retort being sufficient for the engine room alone, while an additional one was required to light up the saloon. The gas was claimed to be made from an oil that did not explode, but to make doubly sure the gas works were situated in a watertight compartment outside the engine room that could presumably be flooded if necessary.

Gas lighting was a clumsy method at best, and the first large-scale electrical installations appeared appropriately enough in the *Umbria* and *Etruria*, the largest and fastest Cunarders of their day, built in 1885. These two ships had four machines each, placed on a lower deck immediately forward of the engine room. They were not very large; three of them were driven by single-cylinder steam engines of 8 inches bore by 6 inches stroke, the fourth of about half the power. These machines supplied current for the arc-lamps in the dining saloon only. By 1890 electric lighting was a general requirement for all large passenger ships, but in the first-class accommodation only, and it had none of the sophistication and good taste of a modern ship's interior. The saloons were illuminated by harsh, flickering arc-lamps while the rest of the accommodation had to make do with sixteen candle-power lamps, the current distribution being by single wire with earth return.

The early steam dynamo was originally belt driven by its steam prime mover and the belt was subsequently replaced by ropes in a grooved pulley. Later still, the dynamo was installed on a rocking cradle and its shaft fitted with a compressed paper pulley which bore against the steam-engine flywheel. By the 1900's the dynamo was direct-coupled to its engine.

The advent of the diesel engine, with its independence of steam, offered the preference of an all-electric engine room to steam-driven auxiliaries, and the choice is still available. There is a certain duality in the motor-ship's engine room where the independent auxiliary machines may be driven either by their individual steam engines or turbines, or else by electric motors, and there is little to choose from, either in economy or reliability, between the two systems. In motor ships with a heavy steam load such as passenger liners or tankers, where one or more boilers are required in any case, the choice is obvious. In most freighters, even where steam is not required, it is customary to fit a donkey boiler to utilise the waste steam from the exhaust gases, and this is sometimes used for power purposes, but the bulk of the power for auxiliary and hotel services is electrical, derived from the oil-engine-driven generators.

The all-electric engine room had one disadvantage however, inasmuch as all the auxiliary machinery serving the main engines shut down upon failure of the electrical supply. Hence a breakdown in the oil engine driving the dynamo would immediately place the vessel at risk, with main engines, steering gear and navigating lights out of action, and it is the rule to have two generating enginse running together when sailing in confined waters such that upon failure of one, the other will take over the essential load.

In the older motor ships with only two auxiliary generating sets, one running and one on stand by, we can appreciate the Chief's worried look when one of his temperamental diesel-generator engines had been pulled adrift for periodical overhaul.

In addition to maintaining the propulsion, lighting and power requirements at sea, the marine engineer also had to care for the refrigerating plant for both catering and cargo purposes, and it is the cargo cooling that will give him most trouble on account of its infinite variety of goods that carry at vastly different temperatures and loading.

Refrigerated cargo was carried experimentally in 1876 between Argentina and France, and in 1880 between Australia and Britain, and by the turn of the century the frozen-meat trade between Britain, Argentina and the colonies was being undertaken by over 150 refrigerated ships. The earliest refrigerating machines worked on the compressed-air system, the cooled air being circulated throughout the cargo chambers and returned to the compressor in a closed cycle.

The machine, usually in duplicate, consisted of a prime mover driving a shaft, which in turn drove the refrigerating unit: an air compressor and an air expander, the latter being just an ordinary steam cylinder with a piston in tandem with the compressor wherein the compressed air expanded to produce work and in consequence lose heat. In addition there were a sea-water-circulated air cooler and a return-air-circulated air dryer. The cargo chamber equipment was simple: charcoal insulated shipsides, bulkheads and deckheads, and a series of delivery- and return-air trunking, usually collapsible for removal and stowage when carrying general cargo.

The air cycle was simple and uncomplicated. Air compressed to 50 lb was delivered to the cooler at roughly 250°F and cooled down to slightly above sea temperature. From there it passed at, say, 60°F through the dryer and lost part of its heat, entering the expander at slighly above freezing point. This air was then expanded through the expander, doing work to help drive the compressor, and in so doing losing a considerable amount of its heat to emerge into the air trunking at 2–4 lb pressure and −80°F temperature. Louvres in the wooden ducting allowed the air to be directed at will throughout the cargo, or enabled the air flow to be reversed.

The compressed-air refrigerating system was popular in cargo liners until the end of the century, mainly because of its simplicity and low pressure, but also because air is non-toxic and free. Its main disadvantage lay in its enormous bulk; its coefficient of performance was only one-sixth of a vapour-compression machine, and a very large compressor was required to cope with the enormous quantities of cold air in circulation. Also, this large mass of cold, moist air deposited snow in the air ducts and snow boxes and these had to be constantly cleared. In addition, temperature regulation of the air circulation was not positive enough to achieve a mild degree of refrigeration: it was frozen cargoes or nothing.

In passing, the air expander in tandem with the compressor was a necessary feature of the refrigerating cycle; if air were merely reduced in pressure from 50 lb to 2 lb without doing any work, the temperature would not drop appreciably. But by reducing this pressure inside a cylinder and making the air put work into a piston, the air would lose heat at the rate of one BThU for every 778 foot-pounds of work delivered.

By the turn of the century, compressed air had been replaced by ammonia as a refrigerant and although this in turn has lost favour to a certain extent on board ship, it is still in use on some German and Russian vessels and, of course, in cold stores.

The ammonia cycle differs from the compressed-air cycle in that it depends for effect upon the latent heat of a gas when it changes its state from vapour to a liquid and back again to vapour. The ammonia gas is delivered from the compressor at about 250 lb pressure and roughly 90°F temperature and condensed in a sea-water-cooled condenser, whence it flows into a liquid receiver, still at 250 lb. From this receiver, the liquid ammonia is released through a regulating valve to evaporate in a brine cooler, dropping in pressure to 20–30 lb and boiling off at a temperature slightly below zero F. The gas is then drawn into the compressor to be recompressed and start the cycle over again. The cooler or evaporator is circulated by brine which in consequence is reduced in temperature and delivered either to a series of air coolers in the holds, where the holds are circulated by cold air, or direct to a number of brine grids covering the shipside insulation and bulkheads. By using brine as a secondary circuit it is possible to achieve very fine limits of temperature control. Occasionally, the ammonia gas is expanded directly in hold grids, the brine circuits being eliminated.

Compared with air, ammonia is a much more efficient refrigerant and its pressures are still in the medium range, compatible with steam pressures afloat. The only serious disadvantage is that ammonia is a dangerous gas to work with in confined spaces. Normal maintenance inevitably requires cylinders and valve chests to be opened out and pipe joints to be broken, and the slightest leak of ammonia into the machinery spaces renders these untenable.

A third refrigerant, the carbonic anhydride or CO_2 gas, came on to the scene somewhat later than ammonia and works on the same principle although with considerably higher pressures, well over 1000 lb. Apart from that, the basic equipment is the same: compressor-condenser-regulating valve-evaporator-compressor, for the primary circuit, with brine circulation between the evaporator and the air coolers or brine grids in the holds as a secondary circuit.

Refrigeration on the CO_2 principle was simple and reasonably safe, and minor leaks passed unnoticed. This attribute, however, constituted a health hazard in itself, as a considerable leak in a confined space, a blown-out gas joint, for instance, could dissolve in the atmosphere and a solution as low as 5% CO_2 could render the air subtly poisonous and lethal. Also, CO_2 was not the perfect gas for shipboard use, especially on the Far Eastern trade routes, as its critical temperature is 88°F. Hence, with sea water in the 90°s, as frequently happens in the tropics, the gas cannot be liquefied and the benefit of latent-heat transfer is lost, with considerable reduction in refrigerating efficiency.

Carbonic anhydride has now gone the way of ammonia for ship's use and has been replaced by a number of related methane and ethane derivatives of the halogenated hydrocarbon family. These products do not have a proper name, they have a designation or a formula not easily recognised (or indeed pronounced) by the layman. To obviate this difficulty they are referred to by standard numbers. Thus CF_2 Cl_2 rejoicing in the appellation of dichlorodifluoromethane, is now known as Refrigerant 12, a useful gas for the smaller refrigerating cargo and domestic plants, working at reasonable pressures, say 150 lb.

Refrigerant 22 has now largely replaced ammonia for the bigger installations on board ship, or for cold stores, etc., having a somewhat higher compression pressure, while Refrigerant 11 has a low-pressure cycle and is useful for high-volume applications such as air conditioning, etc.

Even the compressors themselves are changing in this modern age. The old-fashioned piston drive is gradually being displaced by rotary compressors: screw, vane or centrifugal type, especially on a steady load.

Finally, we come to that most important adjunct of a ship's engine room, the watchkeeping engineer. The very earliest steamboats carried either a mechanic or a millwright, probably the man who built and installed the machinery, and it was he who ran the engine on orders from the bridge. Being a civilian, i.e. not a member of the crew, he was not entered in the Articles and was therefore in the fortunate position of being his own master where the machinery was concerned. By the second generation of steamboats this mechanic had been integrated into the crew.

About mid-century, marine engineers had achieved a form of warrant rank in the Merchant Navy: the owners did not want him but could not afford to do without him, and this was possibly a result of the passing of the 1850 Act requiring Board of Trade certification for masters and mates of foreign-going vessels. The engineer attained officer rank with the introduction of the Act of 1863 demanding certification of engineers. This Act was instrumental in ameliorating engineers' pay and conditions and brought them into line with their opposite numbers on deck, and by the 1890's they had temporarily outstripped their deck colleagues not only financially but in social status.

This period, just before the turn of the century, the era of the mammoth tandem engines, was the heyday of the marine engineer. Years of industrial stability and prosperity had induced in the general public both the urge and the means to see the world, and generations of wealth, if not of breeding, had created a leisured class who demanded immediate service and stood no nonsense from the lower orders: in that unmechanised age, the first-class passengers with true Victorian panache, were not above addressing even that demi-god the Captain as 'their good man'.

But things were different with the marine engineer; he was in the same position, relative to his engines, as the lion tamer in the circus. He it was who controlled the iron monster down below and restrained it from breaking loose and blowing up a boiler or two. He came of a special breed of men for whom this mass of madly whirling machinery held no secrets. These men commanded and received respect from both passengers and crew, a respect duly translated into better pay and conditions.

Now as regards duties, there is not much the watchkeeping engineer does that could not be done equally well by any intelligent passenger after a few hours of coaching. On watch down below the attendant merely has to check temperatures, pressures and levels, and maintain them at a predetermined figure by slight adjustment of the controls; there is nothing in it that requires extensive knowledge or skill, only the ability to withstand a certain amount of discomfort from noise, heat or smell.

It is when things go wrong that the quality of the man shows through: small things like a steam joint blowing out with a stupefying roar or the diesel auxiliary engine failing, and all pumps and steering gear grinding to a full stop in total darkness. Or it could be something a trifle more serious, such as a crankcase explosion or an oil fire in the stokehold. It could even be a complete breakdown of the main machinery through failure of a connecting-rod bolt or a broken crankshaft. The engineer views it all imperturbably; after the smoke and dust have cleared away, he sets about putting things back into working order again. In the case of major breakage, where no spares are available, he carries out a temporary repair to get the ship into port, usually with the minimum of tools: it is his job.

To conclude, a technical magazine of the 1890's sums up the marine engineer's characteristics in a few brief sentences. After listing the engine-room tool-kit in a contemporary tramp: 1 vise-bench, 3 chisels, 6 files, 1 hand and 1 quarter hammer, 36 assorted black bolts, it comments on an emergency repair that brought the ship back to base:

> 'He must be smith, fitter, boilermaker, plumber, etc., as occasion arises. The work he does has no special excellence, you could do it better in your works—so could he. It is the conditions under which the work is done that entitles it to consideration. Leaving the weather—a great factor in the case—out of the question, look at the pluck with which he tackles and the determination with which he carries through with his few tools (often made by himself) repairs that to the uninitiated would appear impossible.'

As someone once put it, the marine engineer's motto is simple and straightforward: 'Difficulties are promptly dealt with, the impossible takes a little longer.'

Conclusion

Well, that is the history of marine engineering: 150 years of development in merchant-ship propulsion that decided the growth and the shape of the ships themselves. From the rickety paddle engine of the 1820's to the stream-lined gas turbine of the 1970's, the marine engine, like a well-bred insect, has undergone a series of metamorphoses to become the fully developed machine of today. The old Gorgon paddle engine was stood on its head to become the screw engine and its simple expansion cylinder increased to quadruple expansion in several cylinders until, at the turn of the century, it had reached its limit in size and weight. Almost overnight it changed into two entirely separate types: the steam turbine with almost limitless potential, and the compression-ignition engine, the nautical equivalent of the workhorse. Although these two dominate the marine field at the moment to the almost total exclusion of all other types, a new and revolutionary engine has appeared on the scene within the last decade and, although we old-time engineers may scoff at gas-turbine drive on the North Atlantic, the recognised proving ground for all new engines, events have already shown us to be wrong. We might, as a profession, recall scoffing at the screw propeller, at compound expansion and at the steam turbine itself.

Keeping step with the development and growth of the engine power, the weight ratio has improved in like proportion. From the early screw engines weighing 600 lb/IHP, machinery weights have steadily improved with each modification and refinement in type and with ever-increasing steam pressures until, at the beginning of the twentieth century the ratio stood at 300 lb/IHP for the average tramp. With the development of new and more sophisticated engines, however, and with the advent of high-pressure turbine techniques, weights came tumbling down to reach 45 lb/SHP in the gas-turbine-driven ship *John Sergeant* and the unbelievable figure of 0·5 lb/HP in the direct-drive gas turbines for hydrofoils.

Similarly with fuel consumption. The ship-owner of the 1820's had to acknowledge 10 lb/IHP/hr as an acceptable figure for his paddle engines, whereas the owner of a jumbo steam tanker expects 0·5 lb/IHP/hr and makes sure of getting it. The engineer's promise of more powerful engines, more economical and very much lighter in weight, have now come true and the marine engine in its own way is rapidly approaching the aircraft engine as a powerpack. The 1820 concept of design and production has rightly given way to the more academic if slower approach of design leading to research, a try-out in development, then on to production, thus ensuring a motive power unit with all the bugs removed and the wrinkles ironed out. But this avenue also leads to overdesign, resulting in a temperamental engine, an engine that needs a very particular brand of fuel and carefully programmed maintenance, and engine maintenance is possibly the most expensive single item in the modern ship.

Again a single-minded pursuit of fuel economy for its own sake makes for a most expensive ship. Marine engines have now reached such a pitch of development that any slight increase in efficiency is gained only at considerable capital cost and with corresponding risk of breakdown. In very general terms, the costs of running a cargo ship can be taken as 50% capital and maintenance charges, 25% for crew's food and wages, and another 25% for fuel. In any complicated machinery, minor saving in fuel costs is often more than offset by extra repair bills. And, of course, maintenance is the ship-owner's biggest worry. Not only does it often incur lengthy periods of demurrage at the most awkward moments, it demands very costly shore facilities and personnel, mostly of the specialist kind.

It is interesting to compare a cargo ship's steam machinery of forty years ago with the present-day motor ship on a similar trade. The old triple-expansion engine with coal-fired boilers was opened out for survey every four years, when piston rings and valves were removed, adjusted and replaced.

Between these four-yearly periods running repairs were carried out on a port-to-port basis by the ship's engineers, and the only spare gear ever placed aboard was packing and jointing and very, very occasionally a set of rings for the feed pump and a dozen or so electric light bulbs. The efficiency and smooth running of the engine room were the responsibility of the engineers; if they failed to keep the engines turning there were plenty of men about waiting for their jobs. In my early days at sea, on tramps, such was the state of the economy that the very few ship's tools were kept locked up in the Second Engineer's cabin. Light slings and strops in both manilla and steel, were spliced in the store during watch keeping, and the stern-tube packing was square-plaited from a ball of spun yarn down the tunnel, each watch taking a spell as a matter of course.

Nowadays diesel engines are repaired by replacement, and in some motor ships I have known there is enough spare gear hung around the bulkheads to build a complete spare engine. In the fast modern cargo vessel only sufficient replacements are carried out at sea to maintain steady-speed conditions, and upon arrival in port the shore gang comes aboard for the machinery overhaul. Although motor ships must pay their way or the ship-owner would be out of business, I cannot support the belief that these very high maintenance costs and inconveniences are recouped by the annual saving in fuel.

Another aspect of the comparison between the old-fashioned cargo ship and the modern freighter, although this is not really my theme, is in the condition of their machinery spaces. The old simple steamer invariably arrived back in its home port with the machinery in showroom condition; gleaming steel, burnished brass and polished copper, and with the chequer plating wiped to a mirror finish. The modern motor ship only too frequently returns home with the engine-room crew exhausted, main and auxiliary engines wheezing visibly and exuding oil at every pore, the ladders, gratings and floor plating foul with carbonised grease and a repair list running to several chapters.

Now, in this history of marine technology I have purposely excluded nuclear propulsion, where steam is generated by a nuclear reactor instead of in the conventional boiler, and where the black gang is replaced by a host of white-coated scientists. The plain truth is that the nuclear-powered ship has no history to speak of, and very little present-day existence, while the future is anything but rosy. To date only four have been built for the world's merchant service, the first being the Russian polar ice-breaker *Lenin*, launched in 1959, and still the most powerful of this type of drive. This ship, of 16,000 tons displacement, is propelled by triple screws all at the stern, the wing shafts developing about 10,000 SHP each, the central shaft 20,000 SHP. Steam at 410 lb is raised in three pressurised-water reactors and is led to four turbo-generators providing power for the three double-armature propulsion motors.

The second atomic powered vessel, the American steamer *Savannah*, was built in 1962, and develops 22,000 SHP on a single screw. The German *Otto Hahn* was laid down shortly afterwards in 1964 though completed only in 1968, a single-screw vessel of a modest 11,000SHP, while the fourth and latest is the Japanese *Mutsu*, in 1969 with only 10,000 SHP on a single shaft. With the *Lenin* on specialised service and the *Savannah* at present laid up, the meagre data available are not sufficient to assess the value of nuclear propulsion and compare it with steam-turbine plant of equal power, although it would seem that economically it would not be viable under 40,000 SHP. This is understandable. The complexity of the reactor requires physicists as well as engineers, and the overall wages bill would probably negative the savings in fuel in anything smaller than a first-class passenger steamer, of which fewer and fewer are being built.

Another feature of modern vessels, automation, has not been touched upon, the argument being similar to that for nuclear power. Centralised and/or automatic controls are a new philosophy in the Merchant Navy, and one that has been received with mixed feelings by owners, crew and underwriters alike. The concept of automation or more correctly cybernation, is as old as machinery itself: engine-speed governors and thermostats, boiler safety valves and feed regulators, they have been installed in ships for generations and are acknowledged to be part of the system. But modern automation means more than that, it controls within strict limits the working temperatures and pressures throughout the engine room and has resulted in ships sailing the high seas whose engine-room crews can all turn in at night safe in the knowledge that the automatic controls and alarms will look after the efficient running of the main and auxiliary engines or, if anything untoward should happen, sound the alarm in the duty engineer's cabin to call him down below.

The underlying theory is quite sound and shipbuilders quite rightly point to the modern motor car whose engine, transmission, electrical and braking systems are a mass of automatic controls and alarms which continue to function in all weathers and with no attention whatever from year's end to year's end. The other side of the case is seldom presented in such glowing colours. Delicate electrical apparatus seldom works efficiently in an oil- or salt-laden atmosphere, and becomes temperamental when exposed to high temperatures, vibration or shock. Also, failure of an automatic component in a motor car would result, at worst, in minor inconvenience; in a ship it could mean an expensive tow home. Engineers of my generation kept our engines turning sweetly by sight, sound, feel, smell and by an extra sense that told us precisely when something was going to happen and ensured that we were on the spot just before it did happen. This might be a pump becoming air-locked, an electric cable smoking, an auxiliary engine running away or an incipient fire in the stokehold. Any of these, if let for a few minutes without attention, might result in disaster. To us the UMS notation (Unattended Machinery Space) smacks of the character in search of a part, very useful in its way and ideal for industrial machinery but not for indiscriminate use at sea, and we would prefer the old system of watchkeeping retained.

But what of the engineers themselves, has life at sea changed very much for the better? The answer is, of necessity, yes and no. Life in tramps in the 1920's and 1930's was pretty grim and not suitable for the gently nurtured. Passages from port to port could take up to two months or more, and once in the tropics, with the ice long since melted in the meat box, it was as well not to enquire too closely into the stew or the liquid, rancid butter. One's mattress consisted of a straw-filled donkey's breakfast and the uniform of the day a pair of khaki duck trousers and a chain-breaker singlet. There was also the time-honoured chore of field days, a two-hour daily spell in addition to watchkeeping, when the winches and windlass were overhauled. This work was not necessarily unpleasant, except on the fore deck in heavy weather, when one just had to keep an eye open for green seas coming over and washing the various bolts, shims and splitpins over the side.

By contrast the machinery, even in motor ships, was simple and uncomplicated and relatively quiet running, while discipline was strictly imposed and recognised; the Chief Engineer's job was a sinecure and was reserved for those who happened to catch Dame Fortune's eye.

In modern cargo ships, living conditions are exceedingly good while pay and leave are more than satisfactory. But all too often, the machinery is of such advanced design and so highly rated that it requires, though seldom receives, very highly trained personnel to look after it. The Chief Engineer of today has far more than the simple routine maintenance to occupy his mind, he has up to a million pounds worth of highly stressed and swiftly rotating machinery under his care, any part of which can fail without warning and bring his ship to an ignominious halt, while he and his crew desperately check through all possible reasons, both electronic and mechanical, to account for the fault.

The old-fashioned marine engines, both steam and motor, were far more considerate in a way, they usually gave ample notice of their intentions before creaking to a halt. Not that they were in any sense trouble-free, they gave the engine-room crowd a great deal of worry and a daily stop to change a main-engine valve was common in my time. But they were so heavily built and ran with such generous clearances that overheating could be checked and corrected before developing into an emergency.

Who among my generation cannot recall taking the Superintendent down below upon arrival at home port and pointing out some of the defects that had been plagueing us throughout the previous voyage? Main boilers whose smoke boxes simply bristled with tube stoppers, or possibly our only good auxiliary engine wheezing compressed air at every joint and labouring just to keep the lights burning? And the confident reply that due to a quick turn-round they would have to do for another voyage and that they would see us home all right.

Somehow they always did!

Index